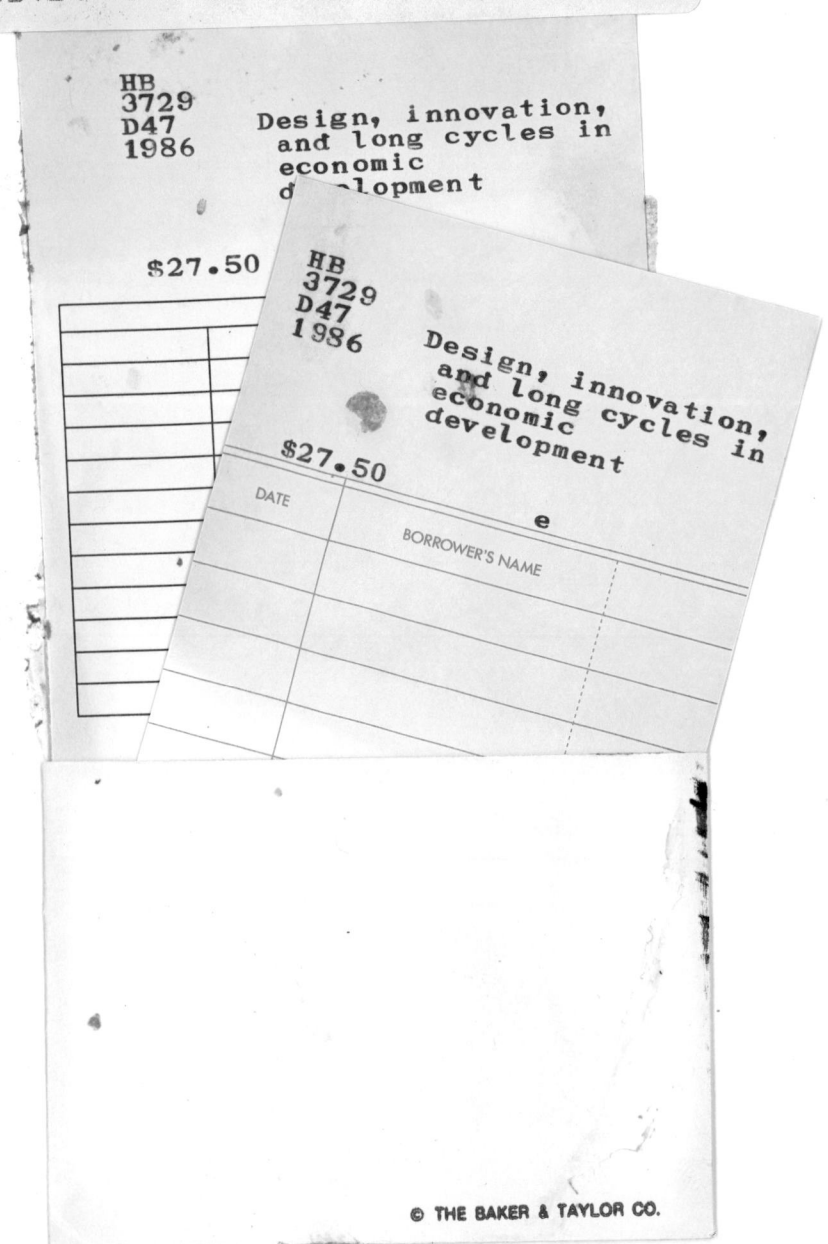

COLLEGE FOR HUMAN SERVICES
LIBRARY
345 HUDSON STREET
NEW YORK, N.Y. 10014

Design, Innovation andLong Cycles In Economic Development

Edited by
Christopher Freeman

St. Martin's Press, New York

© Department of Design Research, Royal College of Art, 1986

All rights reserved. For information, write:
Scholarly & Reference Division,
St. Martin's Press, Inc., 175 Fifth Avenue, New York, NY 10010

First published in the United States of America in 1986

Printed in Great Britain

Library of Congress Cataloging-in-Publication Data
Main entry under title:

Design, innovation, and long cycles in economic
  development.

  Bibliography: p.

  1. Long waves (Economics)--Addresses, essays,
lectures. 2. Technological innovations--Addresses,
essays, lectures. 3. Economic development -- Addresses,
essays, lectures. I. Freeman, Christopher.
HB3729.D47 1986   338.9   85-26272
ISBN 0-312-19449-8

Contents

List of Contributors v

Foreword vii

Introduction 1

1. Technological Innovation and Long waves 5
Nathan Rosenberg and Claudio R. Frischtak

2. Structural Changes and Assimilation of New Technologies in the Economic and Social System 27
Carlota Perez

3. Cycles, Turning Phases and Societal Structures 48
Ger van Roon

4. Technology and Conditions of Macroeconomic Development 60
Giovanni Dosi

5. New Evidence on the Shift Toward Process Innovation During the Long-Wave Upswing 78
Rod Coombs and Alfred Kleinknecht

6. Applying the Biological Evolution Metaphor to Technological Innovation 104
Ugo L. Businaro

7. Design Trajectories for Airplanes and Automobiles 121
J.P. Gardiner

8. Robust and Lean Designs with State of the Art Automotive and Aircraft Examples 143
J.P. Gardiner

9. Long Range Strategic Planning in Japanese R&D 169
Ichizo Yamauchi

10. Marx's Crisis Theory Today 186
Francois Chesnais

11. Long Waves and the International Diffusion of the Automated Labour Process 194
Leonel Corona

12. Long Cycles and the International Diffusion of Technology 214
Luc L.G. Soete

13. The Role of Small Firms in the Emergence of New Technologies 231
Roy Rothwell

Acknowledgements

The work that led to the production of this book was funded by the Wolfson Foundation, and we are grateful to Dr Black, the Director, and to the Trustees of the Foundation for their support.

I am indebted to Christoper Freeman for agreeing to be the first Visiting Wolfson Professor of Design Management, and for providing a structure for the first seminar based on the special issue of 'Futures' in August 1981: 'Technical innovation and long waves in world economic development'.

I am also grateful to all the contributors, many of whom came great distances to attend the seminar, in particular Professor Nathan Rosenberg and M. Francois Chesnais, who provided alternative views and arguments which helped us to focus on specific issues.

I am indebted to Stephen Little for his assistance in the organization of the seminar, and to Charlotte Coudrille, whose expertise and skill in sub-editing and production were invaluable in preparing the work for publication, and to Tim Martin, for his skill in preparing the artwork.

Richard Langdon
London 1984

Contributors

Christopher Freeman
Professor of Science Policy at University of Sussex. Present work at the Science Policy Research Unit on technology policy and long waves. Wolfson Professor of Design Management, Department of Design Research, Royal College of Art, London.

Nathan Rosenberg
Chairman of Department of Economics and Professor of Economics at Stanford University, California, USA. Written widely on the economics of technical change. The authors wish to thank Moses Abramovitz, Paul David, Albert Fishlow, Donald Harris and Sidney Winter for their helpful comments and criticism of earlier drafts. A much condensed version of this paper was published in 'American Economic Review, Papers and Proceedings', May 1983, and a similar version in 'Cambridge Journal of Economics', 8, 1984.

Carlota Perez
Director of Technological Development, Venezuelan Ministry of Industry; visiting Fellow at Science Policy Research Unit, University of Sussex; work on socio-economic impact of technological change, especially as related to Kondratiev long waves. A version of this paper was published in 'Futures', vol.15, no.5, October 1983.

Ger van Roon
Professor of History and Chairman of the Research Group on Long Term Fluctuations, Free University, Amsterdam.

Giovanni Dosi
Research Fellow, Science Policy Research Unit, University of Sussex; Lecturer, University of Venice. Research work on innovation, diffusion and international patterns of trade and production.

Alfred Kleinknecht
Doctoral thesis on long waves at University of Amsterdam; Lecturer in Economics at University of Limburg.

Rod Coombs
Lecturer, University of Manchester Institute of Science and Technology; research on economics of technical change; joint author of forthcoming book on long waves.

Ugo L. Businaro
Research and Development Department, FIAT Company, Turin; representative at EEC, Brussels. A version of this paper was published in 'Futures', vol.15, no.6, December 1983.

Paul Gardiner
Science Policy Research Unit, University of Sussex; research on the economics of design and on industrial diversification policies. Lecturer in Design, Open University.

Ichizo Yamauchi
Industrial analyst with the Nomura Research Institute, London and Tokyo. A version of this paper was published in 'Futures', vol.15, no.5, October 1983.

Francois Chesnais
Consultant, Organization for Economic Cooperation and Development, Science, Technology and Industry Directorate.

Leonel Corona-Trevino
Faculty of Economics, National University of Mexico. Research on industrial automation and effects on productive structure and the international division of labour.

Luc L.G. Soete
Assistant Professor, Department of Economics, Stanford University, California, USA; Senior Research Fellow, Science Policy Research Unit, University of Sussex. Research on technical change and employment, long waves; innovation, diffusion and international patterns of trade and production.

Roy Rothwell
Senior Research Fellow, Science Policy Research Unit, University of Sussex. Research on policies towards technological innovation in industry.

Foreword

Richard Langdon

This collection of papers forms a record of the first of a series of seminars concerned with the management of design and innovation, funded by the Wolfson Foundation.

This first seminar examines the wider context of economic change. The intellectual traditions which came together in this series of seminars may be characterized by the work of Joseph Schumpeter and Herbert Simon. The former is dealt with in detail by Professor Freeman, who puts forward in his introduction to this book Schumpeter's proposition that economic change may be understood in terms of innovations providing the basis for economic expansion. The latter, Herbert Simon, through his connections with the work of the Rand Corporation in the 1960s, provides the link with design. His work, which rests upon the "logic of choice" approach to decision making, drawn from systems analysis and operational research, relies on the specification of needs and objectives. His concern with the dynamics of the design process led him to state in his book, "The Sciences of the Artificial", that "design is concerned with how things ought to be, with the design of artifacts to attain goals", and "everyone designs who devises courses of action aimed at changing existing situations into preferred ones".

This approach has been more widely adopted by design research, where the study of the design process has become a central issue.

Under the guidance of Christopher Freeman as the first Visiting Wolfson Professor, the seminar explored the nature of long cycles in economic development and was intended to provide a wider context and framework for the following seminars in the series. In his inaugural lecture, "Design and British Economic Performance" (delivered at The Design Council, London, 23 March 1983), Professor Freeman had already emphasized the link with the management of research and development.

The series of seminars intends to provide a framework within which to examine how technological progress, economic performance and social development interact and may be assessed and managed. A framework is required which allows for the examination of the management of social and economic policies. Design management must examine the issues related to the creation of a demand for the products of technology and whether this demand is essential to economic growth and social development.

INTRODUCTION

Christopher Freeman

This book contains all the papers given at an international seminar at the Royal College of Art in April 1983. The seminar was itself an indication of the rapidly growing world-wide interest in 'long waves', and since that time there have been many other such indications, notably the conference convened by the International Institute of Applied Systems Analysis (IIASA) in Florence in October 1983.(1)

We have thought it worth while to make these papers available to a wider public, partly because of this growing interest, but also because they have a special emphasis which we believe to be of particular importance in the explanation and understanding of long waves – the role of innovation and design.

There are many different theories of the long wave – some based on long-term trends in capital accumulation and profitability, some on monetary factors, some on industrial relations, some on long swings in the supply and price level for primary commodities. The participants in this seminar also recognized to varying degrees the importance of all or most of these aspects of long cycles of economic development. But they were united in the conviction that one of the central features of any satisfactory long-wave theory must be the relationship between new technologies and the economic and social system.

This should not be taken to imply that any or all of the participants should be regarded as 'true believers' in long-wave theories. As is evident from the papers, some of them were and are sceptical about the idea, notably Nathan Rosenberg. But all of them, including Rosenberg, believed that the topic was of sufficient interest and importance to warrant serious debate and research.

In a sense, the whole seminar could be regarded as a response to Rosenberg's challenge at the outset to produce convincing evidence (a) of the reality of the long-wave phenomenon and (b) of the 'neo-Schumpeterian' explanations of the long waves.

The term 'neo-Schumpeterian' is probably the best shorthand description of the approach taken by most of the seminar participants. Even those such as Rosenberg who remain sceptical about Schumpeter's long-cycle theories would certainly acknowledge Schumpeter's seminal contribution to the whole debate about technology and long-term trends in economic development. Moreover, they would also agree that many aspects of this complex relationship cannot be encompassed by short-term business cycle theory or product cycle theories which ignore the long-term characteristics of many 'technological trajectories' and diffusion processes, and the interconnections between social, organizational and technical innovations. Paul Gardiner's paper is of particular interest in its demonstration of the connection between long-term technological trajectories and the 'robustness' of designs.

However, although the flavour of most of the contributions could reasonably be described as neo-Schumpeterian, this does not mean that they subscribed to any attempt to establish a new orthodoxy based on his ideas. One of the most important and stimulating papers at the seminar – that of Carlota Perez – took as its starting point the failure of Schumpeter to develop any satisfactory theory of depression. Van Roon also commented on the dearth of research on the lower turning points of long waves.

Carlota Perez points out that although Schumpeter had a reasonably adequate explanation of the upper turning point in the long cycles, he had no such explanation for the lower turning points. In her own theory she suggests that deep depressions should be regarded as periods of 'mis-match' between the new technological paradigm which has emerged and the institutional and social framework within which it is diffusing. Whereas technical change at the micro level can proceed extremely rapidly, social institutions at the macro level are characterized by a much higher degree of inertia and indeed by efforts to resist change, based on the interests of groups associated with older industries and technologies, now threatened with decline or even disappearance. Depressions therefore should be seen as periods when there is an intensive search process for new social and political solutions to problems of institutional adaptation. Each new technological paradigm has characteristics which may differ fundamentally from earlier dominant technologies in terms of skill requirements, management attitudes and strategies, firm structures, infra-structural investments, government policies and so forth. Only when the search process yields workable social solutions can a new coherence be established between the new technologies to be realized in terms of productivity, profits and prosperity.

Yamauchi's paper on long-term planning in Japanese research and development provided an extremeley thought-provoking commentary on trends in Japanese policies for technology, which fits well into the Perez framework since it suggests the possibility that the process of institutional adaptation may have already gone quite a long way there, so that Japan could take over the leadership in the fifth Kondratiev upswing, rather in the way that Germany and the United States took over world technological and economic leadership from Britain in the upswing of the third Kondratiev.

The fruitful concept of relative inertia in the social system as compared with periods of rapid change in technology is also taken up by Giovanni Dosi, who extends the notion of differential rates of change to science as well as technology. He points out that both fundamental science and technology are driven, at least in part, by semi-autonomous internal mechanisms; this creates in principle the possibility of recurring periods of mis-match associated with paradigm changes.

The theme of continuity and discontinuity in the development of technologies was also a feature of several of the other seminar papers – notably those by Businaro and Gardiner. Businaro explained some of the advantages and limitations of the analogy with biological evolutionary processes, whilst Gardiner provided a very important link between the theory of technical innovations and the theory and practice of design.

One of the positive achievements of the seminar was to explore and develop this link, which is a central interest of the Department of Design Research at the Royal College of Art. From the purely British point of view, the links between design and innovation are a critical area of the relatively weak post-war British economic performance in international trade. These specifically British problems were not the main concern of the seminar, which was an international one, but they were taken up by Richard Langdon, (2) and in my own lecture for the RCA, 'Design and British Economic Performance'. (3)

From the international point of view, the interesting problems relate to the changing location of international technological and economic leadership in successive Kondratiev waves and the rate at which other countries can catch up with technological leaders. The possibility (some would say probability) that Japan is taking over the leadership for the next upswing of the world economy has already been alluded to. Luc Soete discussed the more general problem of the way in which technology gaps are first opened up and later closed by overtakers and followers during the course of a long cycle. Both he and Corona point to the special problems of Third World countries in this connection. Soete's paper is an important first step towards a general theory of the international diffusion of technology, following his major contribution to neo-technology theories of international trade. (4) The participation of scholars from the Third World was a particularly welcome aspect of this seminar, but the role of the Third World in long cycles and the relationship between Schumpeterian theories and those of Rostow remain a relatively unexplored territory.

How far the seminar succeeded in responding to Rosenberg's challenge is a matter for the reader to judge. Obviously, it is not a challenge which can be met in a matter of a few days, but in my own view the seminar did mark an important advance in the development and consolidation of the neo- Schumpeterian position in four respects.

First of all, it went well beyond the older controversies surrounding the original Mensch (5) theories of clustering of innovations. The papers by Coombs and Kleinknecht and by van Duijn indicate the extent to which a fruitful convergence of ideas is developing in this area. That of Carlota Perez demonstrated a new way of conceptualizing the whole process with its emphasis on the crystallization of a new technological paradigm during the previous Kondratiev wave, when the older paradigm was reaching various limits to its further growth and was becoming increasingly characterized by diminishing returns and a pattern of incremental innovation.

Secondly, this greater clarity in relation to the clustering of innovations and new technological paradigms was also reflected in the discussion of structural change in industry. Recognition of the bipolar nature of the distribution of innovative effort also has relevance to the role of large and small firms during the course of the long cycle. Although the overall tendency of the economy has been towards increasing concentration in the last few decades, there has been a counter-currentofr the emergence of new small firms in some high-technology areas. Rothwell's paper indicates the importance of this recognition and the dangers of oversimplifying the unevenness and complexity of structural change in industry.

Thirdly, the seminar did demonstrate conclusively the relevance and importance of relating the international diffusion of technology to the pattern of long cycles.

Finally, through achieving greater clarity on the relationship between the changes in science, technology and design and those in the structure of industry and the institutional and social framework, the seminar did lay the basis for the development of appropriate policies to handle the problems of social change now confronting all of us at both national and international level. However, van Duijn's closing paper showed admirably how serious the present problems are, and how far we have to go before we are in a position to influence the course of events in a more positive direction.

Notes
1. Bruckmann, G. and Vasco, T. (1983) 'Long Waves, Depression and Innovation', Proceedings of the International Institute for Applied Systems Analysis (IIASA) Workshop in Siena/Florence, 26-29 October 1983.
2. Langdon, R. (1984) 'Design and Industry', vol. III, Design Policy Conference Proceedings, Design Council.
3. Freeman, C. (1983) 'Design and British Economic Performance', paper presented at the Design Centre as part of a Visiting Wolfson Professorship of Design Management at the Department of Design Research, Royal College of Art, London.
4. Soete, L.L.G. (1981), A General Test of Technological Gap Trade Theory, 'Weltwirtschaftliches Archiv', Band 117, Heft 4, pp.638-660.
5. Mensch, G. (1975) 'Das Technologische Patt', Frankfurt, Umschau (English edition 'The Technological Stalemate', MIT, 1979).

1 Technological innovation and long waves

Nathan Rosenberg and Claudio R. Frischtak

This paper is about the existence of long cycles or long waves of economic growth. No one who has examined the dynamics of capitalist economies over long historical periods can doubt that they experience significant long-term variations in their aggregate performance. The question is whether these long-term variations are more than the outcome of a summation of random events, and further, whether they exhibit recurrent temporal regularities that are sufficiently well behaved to call them "long waves". In recent years there has been a strong resurgence of interest in such long-term movements, since their existence could provide a coherent explanation for the poor performance of capitalist economies over the past decade.(1) This renewed interest also reflects a search for alternative ways of explaining the unbalanced nature of the growth processes of mature capitalist economies that go uncaptured by the Solow-Swan paradigm, in its concern with equilibrium dynamics and steady states of one, two or multi-sector representations of the economy.

The study of price and output swings of extended duration has a long tradition, having initially drawn the interest of both Marxist and non-Marxist writers round the turn of the century.(2) Yet, it was the work of Kondratiev(3) which constituted the first systematic attempt to confirm such movements with data that included not only prices, interest rates and wage series, but foreign trade, industrial production and consumption for France, Britain, and (to a lesser extent) the USA. Kondratiev concluded that the data suggested the existence of long cycles with an average length of 50 years, and going back to the end of the 18th century. However, in formulating the possibility of long cycles, Kondratiev expressed himself with great caution, calling attention to the fact that the available historical evidence dealt with fewer than three full cycles as well as to the poor quality of production time series before 1850.(4)

Kondratiev's hypothesis gave rise to two distinct lines of historical research, one centred around the notion of a price (or interest rate) swing, and another focussing on long waves as a phenomenon in real quantities. Taken as price swings, long waves have been alternatively interpreted as the outcome of real and monetary forces; whereas taken as fluctuations in real quantities, long waves have been understood as ultimately driven either by the process of capital accumulation, as in Kondratiev,(5) or by that of technological innovation, as in the Schumpeterian tradition.

This paper is not an attempt to examine the historical evidence for long cycles. We have in fact examined this evidence and we find it unconvincing. Although historical data might conceivably lend some plausibility to the notion of long cycles in prices, we remain, at present, sceptical of the case that has so far been made for their presence in real phenomena - i.e. in aggregate output or employment (moreover, even with respect to prices, it is very unlikely that there have been long waves during the upward drift in absolute prices of the last half century, although there have certainly been drastic changes in the terms of trade between industrial goods and primary commodities).(6)

What we offer here instead, and consistent with our present scepticism, is an attempt to examine the economic logic of long waves. More specifically, we ask: what conditions would need to be fulfilled in order for technological innovation to generate long cycles in economic growth of the periodicity postulated by Kondratiev and his disciples? Surprisingly, in view of the amount of current interest in the subject, this question is hardly ever addressed by advocates of long waves with sufficient analytical rigor. In fact, an adequate or even a plausible theory of long cycles, based primarily upon technological determinants, does not presently exist, although belief in such long waves is now widespread.(7) It is our view that such a theory, which might account for the presence of long cycles in some real economic variable, would have to fulfil a set of logically interdependent requirements. We discuss these requirements under the four categories of causality, timing, economy-wide repercussions and recurrence.

1. Causality

The first of the requirements for a technological theory of long cycles is a clear specification of causality among the factors associated with this phenomenon. Kondratiev was insistent that capitalism had its own internal regulating mechanisms, and he regarded the pace or rhythm of the long cycle as an expression of these internal forces. Long cycles, as Kondratiev put it, "arise out of causes which are inherent in the essence of the capitalist economy" (8). The cyclical behaviour of the capitalist economy in turn shapes the conditions that are favourable to technological innovation. In this specific sense, therefore, technological activities stand in the position of dependent variables whose volume and timing are determined by those deeper-rooted forces that shape the rhythm of capitalist development.(9)

Furthermore, Kondratiev sees an unusually wide range of economic and social phenomena as being endogenously shaped - not only technological innovation but also wars, gold discoveries, and the entry of new geographical regions into the nexus of market relationships with capitalist economies. It remains true, of course, that Kondratiev views technological change as exercising an important influence on the course of capitalist development; yet the essential point is that these technological changes are viewed as occurring in response to endogenous forces within capitalism.(10)

Schumpeter was, of course, the foremost and most influential articulator of the opposite view - that long cycles are caused by, and are an incident of, the innovation process. Indeed, Kondratiev's ideas were first brought to the attention of English-speaking economists through Schumpeter's treatise on business cycles, in spite of the fact that Schumpeter urged a causality that was sharply in contrast with Kondratiev's. Moreover, it is the Schumpeterian variant of the long cycles hypothesis, stressing the initiating role of innovations, that commands the widest attention today.

In Schumpeter's view, technological innovation is at the centre of both cyclical instability and economic growth, with the direction of causality moving clearly from fluctuations in innovation to fluctuations in investment and from that to cycles in economic growth.(11) Moreover, Schumpeter sees innovations as clustering around certain points in time - periods that he referred to as "neighbourhoods of equilibrium", when entrepreneurial perception of risk and returns warranted innovative commitments.(12) These clusterings, in turn, lead to long cycles by generating periods of acceleration (and eventual deceleration) in aggregate growth rates.(13) Just why clustering should occur is obviously crucial to a theory of long cycles and we will therefore return to this question shortly. But it is essential to stress that an exponent of the view that technological change is at the root of the long cycle needs to demonstrate (a) that changes in the rate of innovation govern changes in the rate of new investment, and (b) that the combined impact of innovation clusters takes the form of fluctuations in aggregate output or employment.

Such causal links are not demonstrated in the neo-Schumpeterian literature. Consider the work of Freeman et al.,(14) which we regard as the best statement on long waves from this perspective. The conditions which set in motion the diffusion and clustering of basic and related innovations and which would stand behind the upswing of a long wave are only loosely specified. The authors variously stress the "role of advances in basic sciences, and social, managerial and organizational changes in triggering and facilitating clusters of basic inventions and ninnovations",(15) "the phase of the long wave, particular breakthroughs in technology" and their "natural trajectories" as influences in the clustering processes,(16) "the scientific, technical and economic links" among constellations of widely adaptable innovations,(17) and "a social change which permits a market to grow rapidly or large amounts of capital to be raised or invested".(18)

In spite of this long listing of possible influences, we are left without a precise knowledge of what are the necessary and sufficient changes in the environment which, even conceptually, can bring out a bandwagon-like diffusion of some number of basic innovations. In other words, there is no well-specified set of elements that effectively link and elucidate the direction of causality between the basic innovations, the "general level of profitability and business expectations", and their diffusion in the form of a swarm of new products and processes. More generally, nowhere in the literature is there to be found an unambiguous treatment of causality, within a neo-Schumpeterian framework, which establishes the precedence of innovation clusters over investment outlays and aggregate movements in the economy.

2. Timing

The process of technological innovation involves extremely complex relations among a set of key variables - inventions, innovations, diffusion paths and investment activity. The impact of technological innovation on aggregate output is mediated through a succession of relationships that have yet to be systematically explored in the context of long waves. Specifically, the manner in which various economic and technological forces may influence the lag between invention and innovation, as well as the speed of the diffusion process and the impact on output growth, are insufficiently appreciated.(19) A technological theory of long cycles needs to demonstrate that these variables interact in a manner that is compatible with the peculiar timing requirements of such cycles.(20)

It is not enough to argue that the introduction of new technologies generates cyclical instability. It is necessary to demonstrate why technological innovation leads to cycles of four and a half to six decades in length, with long periods of expansion giving way to similarly extended periods of stagnation.(21) Of course the burden of establishing such a connection lies with proponents of the long cycle theory. We therefore confine ourselves here to a brief inventory of strategic factors that may be expected to determine the length of the time period required for the introduction of new technologies and the realization of their full impact upon aggregate output. In particular, long waves involve a diffusion period of appropriate length, the spacing (non-overlapping) of substitute technologies, and the clustering of those which are of complementary and unrelated natures.

New inventions are typically very primitive at the time of their birth. Their performance is usually poor, compared to existing (alternative) technologies as well as to their future performance. Moreover, the cost of production, at this initial stage, is likely to be very high -

indeed, in some cases a production technology may simply not yet exist, as is often observed in major chemical inventions (nylon, rayon). Thus, the speed with which inventions are transformed into innovations, and consequently diffused, will depend upon the actual and upon the expected trajectory of performance improvement and cost reduction.

This process is rendered more complex, first by the fact that in the early stages, when performance is still very modest and production costs are high, improvements leading even to significant cost reductions may have no sizable effect upon rates of adoption. When, on the other hand, the new product attains cost levels roughly equivalent to those prevailing under the older technology, even small further cost reductions may lead to widespread adoption. Or, alternatively, at this point relatively small changes in factor prices may shift the balance sharply in favour of the innovation, depending upon the nature of its factor-saving bias. Thus there may be a highly non-linear relationship between rates of improvement in a new product and rates of adoption. Further, there is often a long gestation period in the development of a new technology during which gradual improvements are not exploited because the costs of the new technology are still substantially in excess of the old. However, as the threshold level is approached and pierced, adoption rates of the new technology become increasingly sensitive to further improvements.

Second, since innovation and investment decisions are future-orientated and therefore inevitably involve a high degree of uncertainty, adoption and diffusion rates are also powerfully shaped by expectation patterns. In certain cases, these expectational patterns may lead to a prolonged delay in the introduction of potentially superior new technologies by adversely affecting their expected profitability. This could be the result of uncertainty regarding the timing and significance of future improvements in the technology being considered for adoption; of the expected availability of substitute innovations; and of the expected and actual improvements along the old trajectory.(22) Indeed, it has been very common for competitive pressures generated by a new technology to lead to substantial improvements in the old technology, so that the new one establishes its superiority more slowly than would otherwise have been the case.

Note in this respect that major improvements in productivity, as well as output growth, may be stretched out over very long time periods, as a product goes through innumerable minor modifications and alterations in design. The camera, a mid-19th-century innovation, experienced remarkably rapid diffusion in the post-World War II years. The Fourdrinier papermaking machine was patented in England and France in 1799. The first machine was made in 1803. In spite of innumerable modifications, its basic principles of operation remain the same, and the machine continues, more than one and three-quarter centuries later, as the dominant technology in the manufacture of paper. Substantial productivity improvements continue to be developed within this technological framework. More widely used products like a steam engine, electric motor or machine tool have experienced a proliferation of changes, as they are adapted to the diversity of needs of numerous ultimate users.

To the extent that major innovations vary relative to the time period for which they remain important, in part because substantial improvements will often take place long after the initial introduction of the innovation, it renders highly problematical the whole exercise of inferring a Kondratiev long cycle from a particular innovation. How does one date the long cycle associated with the steam engine? Beginning with Watt's seminal inventions in the 1770s? What we know about the slow pace of its adoption in the late 18th century renders this extremely doubtful. But, in addition, the improvements associated with the compound engine brought huge productivity improvements sufficient to introduce the steam engine to important new uses - and this came a full century after Watt's major contribution. How does one date the impact of the airplane? It was about thirty years after the first successful achievement of the Wright Brothers at Kitty Hawk that the airplane had a significant commercial impact - with the introduction of the DC-3 in the mid-1930s. But with the subsequent innovation of the jet engine, fully a half century after the first achievement of lighter-than-air flight, the commercial impact of the aircraft increased by at least an order of magnitude.

One may say, of course, that the compound engine and the jet engine each deserve to be treated as separate innovations in themselves, with contributions to separate long cycles. There is nothing objectionable, in principle, about such a procedure, if it can indeed be employed to provide a convincing account of historical change involving long waves.(23) For the present we only wish to assert that such an accounting remains to be presented by advocates of technologically determined long waves.

Third, the adoption of a new technology is often critically dependent upon the availability of complementary inputs or, in some cases, upon an entire supporting infrastructure. Automobiles required extensive networks of roads, petrol stations and repair facilities. The electric lamp required an extensive system for the generation and distribution of electric power. Seldom do new products fit into the existing social system without some intervening period of accommodation during which these complementary considerations are arranged. Not only does this signify a heavy commitment to an established technology, and a further reason for a slow initial shift to a best practice frontier but, moreover, the time period required for such accommodations may vary greatly from one innovation to another.

Even if major innovations experience appropriately long and logistically shaped diffusion paths with technologies going through phases of accelerated growth and eventually petering out, it does not necessarily imply that the slack in the declining phase of an individual innovation cycle might not be taken up by other technological innovations, thus eliminating the impact of a long phase of 'sectoral' retardation. What would still be needed for a wave-like pattern of growth is that other major substitute innovations were excluded until the original one had run its course. Without such a spacing mechanism, partially overlapping innovations might otherwise generate steady rates of growth rather than cycles.

What technological forces might impose cyclical behaviour rather than some sort of relative stability of economic activity along any given path traced out by a sequence of major substitute innovations? We have already suggested three such forces that might delay the introduction and widespread adoption of a new substitute technology, namely, a production cost differential that may still persist between the old and new technologies; certain expectational patterns held in common by entrepreneurs regarding improvements in both technologies; and the costs associated with scrapping and replacing the infrastructure committed to the old one. An additional possibility is that major innovations may establish certain trajectories of readily-available performance improvements and cost reduction (more circuits on a chip, fewer pounds of coal per kilowatt hour of electricity). Engineers and technically trained personnel often work with such implicit notions.

Thus, the awareness of these trajectories may serve as focusing devices that fix the attention of engineers upon teasing out the further improvements that are understood to be available from the existing technological framework, rather than searching for entirely new technologies. In this sense, the commitment of large amounts of resources to the exploitation of electric power may have been retarded so long as there seemed to be high payoffs available from further improvements in the technology of steam power, just as the nuclear power option may have been seriously explored only when there was a growing sense that further improvements in fossil fuel burning installations were approaching exhaustion.

In addition, these trajectories may be expected to shape the educational system and the training of engineers and other technical personnel. The inertial forces here may strengthen the commitment to an existing technology and render more difficult the exploration of new realms of technical possibilities. Whether such trajectories in fact played an important role in the spacing of technological innovations is an interesting hypothesis on which there is, at present, little evidence. Proponents of long cycles might find it worth exploring, because it may serve as an important support of the long-cycle hypothesis.

The reasons so far invoked for lengthy delays in the adoption of new technologies, of a kind that might produce extended periods of industry-wide stagnation, were discussed in connection with major substitute innovations. Should similar considerations of spacing be extended to cases of unrelated or complementary technologies unfolding along many different trajectories?

In the case of unrelated technologies, the answer, prima facie, would be no. Here it is important to distinguish between the impact of innovations that compete with existing technologies in a given industry or sector, and those that do not. Even if one argued that there were forces leading to the spacing of innovations in the same industrial sector, in the sense that the arrival of a new technology has to wait until the benefits of moving along the previous technological trajectory had been largely exhausted, this would be of limited relevance for major innovations in other industries. The fact that we are still on a highly productive portion of the steam trajectory might conceivably tell us something about the timing of substitute innovations, such as electric motors, but little, if anything, about the timing of unrelated innovations and their subsequent diffusion in electronics, synthetic fibres or pharmaceuticals.

Yet the long-cycle hypothesis might be considerably strengthened if a large number of unrelated innovations had the main phases of their life-cycles synchronized by macroeconomic conditions. Indeed, the simultaneous diffusion of a large number of unrelated innovations may be understood as being regulated by general conditions in financial, factor and product markets. If favourable, they might lead to a 'bandwagon' effect along a number of separate industry trajectories, followed eventually by a slowdown.(24) The result would be an innovation cluster of type 'M', the vertical summation of sectoral logistics. Its impact would take the form of a period of fast and then slow growth rates, but now in many different trajectories simultaneously.

In the case of related technologies, an additional reason can be invoked for the synchronization of separate diffusion paths: they may be linked by a system of technologically connected 'families' of innovations, made up of complementary, induced and closely related ones.(25) This would come about because the interactions of a few basic technologies would provide the esential foundations for other technological changes in a series of ever-widening concentric circles. A technological cluster, or a cluster of type 'T', arises therefore when one (or a small number of) major related innovations provide the basis around which a large number of further cumulative improvements are positioned. Let us look in more detail for the technological reasons why innovations come in clusters, and not in a continuous stream.

1. Innovations breed other innovations because one innovation may sharply raise the economic payoff to the introduction of another,

Technological innovation and long waves

bringing those which are known to be technically feasible but so far economically unattractive to the point of adoption. Moreover, there are internal pressures within a technological system which serve to provide inducement mechanisms of a dynamic sort. The attention and effort of skilled engineering personnel are forcefully focussed upon specific problems by the shifting succession of bottlenecks which emerge as output expands. More generally, an innovation leads to further innovations to the extent that it provides a framework that makes it possible to conceptualize, design and work on a number of complementary and related technologies.(26)

2. Innovations breed investment in so far as new products and processes call for new vintages of machinery and equipment and, as we previously mentioned, the availability of a complementary infrastructure. Yet the reverse is also true: investment stimulates inventive and innovative activity. It is an incentive for inventions, if they are understood as an economic response to shifting financial payoffs to their different categories.(27) Perhaps more important, infrastructure investment, once in place, serves as inducement to the introduction and adoption of innovations which plug in to the already existing supportive apparatus. Once an extensive electric power distribution system has been installed to meet the requirements of residential lighting, the time required to achieve high rates of adoption for a variety of other electricity-using consumer durables is considerably reduced.

The technological clustering of innovations around different diffusion paths should therefore be taken as a corollary of the fact that certain innovations bring other innovations directly, by providing a working frame within which further innovations become possible, and indirectly, either through their forward investment links, or the backward connections with inventive activity.

We have shown so far that:

(i) Technological forces exist which may lead to cyclical behaviour in certain industries, where major innovations come to substitute for one another sequentially in time.
(ii) There also appear to be technological reasons for industries which stand in a complementary relationship to each other to experience common fluctuations in economic activity (beyond the more obvious technical complementarities in production).
(iii) There are macroeconomic reasons for apparently unrelated industries to have the pace of their economic activity synchronized over time.

Yet the basic question persists: is spacing within and synchronization among different diffusion paths, for both technological and macroeconomic reasons, sufficient to provide a long-wave pattern of aggregative growth? And if so, how?

It is our present tentative assessment that modes of argument at the technological level, while potentially interesting and well worth further exploration, will be of only limited usefulness in providing a convincing account of the generation of long waves. The mechanisms within the purely technological realm appear to be insufficiently robust for this purpose. Technologically driven long waves can be made to appear plausible only if macroeconomic factors can be shown to play a dominant role in shaping and disciplining the timing of the introduction of innovations. The beginning of an upswing would therefore be characterized by a sufficiently large stimulus from the 'M' clustering process upon the previously positioned 'T' cluster. In other words, at the initial phase of industry life-cycles, the state of the economy (or 'market conditions') regulates, to a large extent, the still incipient sectoral demands, so that the introduction of new products and processes tends not to occur unless the economic environment is conducive to increases in consumer spending and investment activity. On the other hand, once activated by the macroeconomic environment, the technological long cycle is required to detach itself from swings in demand which closely track short run changes in macroeconomic conditions and instead follow the internal dynamics of technological factors. Such autonomy might indeed be observed once the new industries have surpassed their initial (experimental) phase and before reaching maturity (when output changes are again in line with aggregate demand changes in the economy). At that point, the dissemination of new products and processes would present a self-sustaining mechanism, not only to the extent that such innovations tend to cluster in a pattern of mutual feedback and reinforcement, but moreover because they actively substitute and displace older products and processes. In sum, they create a market for themselves in direct relation to their scope for substituting for mature commodities and complementing new ones.

The outcome of this non-trivial interaction between factors which belong to the technological realm and those which are reponsible for decisions to carry out substitution in consumption and production may lead, in a manner which we have previously discussed, to extended periods of multisectoral growth and retardation, although there is no reason to believe that they will add up to cycles of 45 to 60 year duration. What has not been shown so far is the connection between such factors and the derived and induced demands for capital and consumer goods which would account for the economy-wide impact of innovation clusters.

Therefore, to argue effectively for a technologically driven long cycle, an additional requirement has to be met: the cluster of innovations must occupy a strategic position in the economy in terms of backward and forward links, the subject of the next section.

3. Economy-wide repercussions

An essential step in a technological theory of long cycles is the demonstration of the mechanisms through which particular changes in technology exercise sizable changes in the performance of the macroeconomy. Much of the present literature on long cycles cites specific innovations in association with specific historical long cycles, but without even attempting to demonstrate how these innovations, or innovation clusters, might be expected to exercise macroeffects of the size required by a long cycle model. Precise quantification is admittedly an impossible task partly, as we will see, because of the elusive nature of the relationships, but also because of the data requirements that such an exercise would involve. Nevertheless, to be persuasive, or even plausible, some estimates of at least the orders of magnitude involved are indispensable.

The economy-wide impact of technological innovations needs to be understood not only in terms of the direct impact of cost reductions and the release of resources to alternative uses, but of the strength of their backward and forward linkages.(28).

1. They should be strongly linked backwards in terms of expenditures for building, machinery, equipment and raw materials, such that the initial innovation and investment requirements lead to further investment decisions in the production goods sector. Historically, this second wave of investment has often bred a second wave of innovations, more explicitly 'process' oriented, and concentrated in the production goods sector. It should be particularly noted that this last set of innovations has frequently had the effect of increasing the productivity of the economy at locations far from the specific sector that originally gave rise to the innovative activity.

2. The impact of innovations will also depend upon the strength of their forward linkages. These might take the form of a reduction in the price of the products into which the innovation enters as an input, leading to an expansion in the size of their market, and therefore also to an expansion in the rate of capital accumulation, output growth and technical progress in these industries. These induced responses would depend on the number of industries into which the innovation enters as an input, its substitutability for other inputs, the proportion of total costs it accounts for and the extent of cost reductions it imposes upon the product.(29) More important, innovations may induce the creation and diffusion of new products and processes that, in their turn, would bring about the widespread adoption of the original innovation (the microchip is a compelling recent example). Alternatively, the impact will depend upon the extent to which the initial innovation proves to be at the core of 'major natural trajectories' (such as the electric motor in relation to the process of electrification), or more generally, in key sectors of the economy, such as energy and

transport.

It is therefore the strategic location that innovations occupy within the economy, in terms of generating both investment and further technological change, which may tell us a great deal about their ability to generate a long cycle growth pattern.(30) Yet this is particularly difficult to assess, as it is characteristic of technological innovation that it leads to wholly new patterns of specialization both by firm and by industry, with the result that it is impossible to compartmentalize the consequences within conventional Marshallian industry boundaries, or to read their impact directly off an input-output table.(31)

The ways in which technological changes coming from one industry constitute sources of technological progress and productivity growth in other industries defy easy summary or categorization. In some cases relatively stable relationships have emerged between an industry and its supplier of capital goods, and that becomes of decisive importance for the rate and direction of technical change in that industry, as evidenced by the post-war experience of the American aluminium producers. On many occasions the availability of new and superior metals or new alloys has played a major role in bringing performance and productivity improvements to a wide range of industries, as in the cases of railroads, machine tools, electric power generation, and jet engines, among others.

Often an innovation from outside will not merely reduce the price of the product in the receiving industry, but will make possible wholly new or drastically improved products or processes. In such circumstances it becomes extremely difficult even to suggest reasonable measures of the impact of the triggering innovation, because such innovations, in effect, open the door for entirely new economic opportunities and become the basis for extensive industrial expansion elsewhere. In the 20th century the chemical industry exercised a massive effect upon textiles, at the time a very 'mature' industry, through the introduction of an entirely new class of materials - synthetic fibres. Technological change in the chemical industry has exercised a similar triggering function in other industries than textiles. In the case of the electrical industry, the chemical industry played a critical role through the provision of refractory materials, insulators, lubricants, and coatings, and provided metals of a high degree of purity for use in conductors. And yet, the profound effects of chemical innovations have had a relatively limited visibility because of the intermediate good nature of most chemical products.

In sum, the inter-industry flow of new materials, components and equipment may generate widespread product improvement and cost reduction throughout the economy. This has clearly been the case in the past among a small group of producer goods industries - machine tools, chemicals, electrical and electronic equipment. Industrial purchasers of such producer goods experienced considerable product and process improvement without necessarily undertaking any research expenditure of their own. Such inter-industry flow of technology is one of the most distinctive characteristics of advanced capitalist societies, where innovations flowing from a few industries may be responsible for generating a vastly disproportionate amount of technological change, productivity improvement and output growth in the economy. It is certainly conceivable that technological change generates long cycles through such inter-industry flows and their consequent macro-economic effects. Yet, given the difficulties of knowing what is the nature of the benefits flowing from each innovation, and where exactly within the structure of the economy these benefits eventually accrue, this can at best be regarded as no more than an untested hypothesis until systematic attempts at quantification have been undertaken.

4. Recurrence

The final requirement for a theory of long waves based upon technological innovations involves demonstrating their cyclical or recurrent character. In fact, it is not sufficient to show that causality runs from innovation to investment; that the economic and technological factors which determine the adoption of new technologies do so in a manner compatible with the stringent timing requirements of a Kondratiev; and that the patterns of diffusion and inter-industry linkage of new technologies involve sufficient amplitude that long cycles are perceived in the form of sizable variance in aggregative growth rates. It still has to be shown, if the argument is going to be logically complete, that the waves repeat themselves over time, either because the wave-generating factors in the form of innovation clusters are themselves cyclical (or at least recur with a certain regularity), or because there is an endogenous mechanism in the economic system which necessarily and regularly brings a succession of turning points.

What are the conditions under which long cycles become a historical necessity, in the sense that there are structural reasons for one long wave to follow another?

(i) the availability of an elastic supply of inventions, at a time when risk-return combinations appear propitious for innovations;

(ii) the formation of a cluster of innovations at the base of the upswing, i.e. a technologically dense set which undergoes a rapid process of diffusion under favourable macroeconomic conditions;

(iii) the reaching of an upper turning point of the technologically-driven cycle due to increasing macro-economic instability, as well as forces that deter the introduction of substitute technologies;

(iv) the arrival of the economy at a technologically fertile ground, after an appropriately extended period of time. At this point, old innovation paths have been largely exhausted, but previously postponed ones have not yet been taken up.

This schema brings numerous problems. One would be hard pressed to show that Kondratievs are regulated on a purely internal basis, and that, in the past, exogenous factors have had only a marginal effect upon such long-term improvements. Others have in fact argued that the recurrence of innovation clusters has been more in the nature of historical accidents than endogenously generated fluctuations in the rate of innovation.(32) Moreover, our earlier discussion of timing provides no compelling reason to expect recurrence at 45- to 60-year intervals, even if innovations cluster and such clusters appear regularly.(33)

Furthermore, in view of the widely varying lengths of individual innovation cycles, one cannot be certain, in observing a Kondratiev upswing, that the upswing is a consequence of a recent innovation or innovation bunch, rather than the protracted acting out of the longer-term impact of earlier innovations. We suspect that this is a much more serious problem than is generally recognized in recent attempts to associate specific cyclical upswings with specific antecedent innovations. At any time, the investment opportunities generated by specifically recent innovations are likely to be small compared to the aggregate of opportunities made available by ongoing improvements and modifications in older technologies. Moreover, much empirical evidence suggests that many innovations (such as the car) play a major role in more than one Kondratiev. As a result, we suspect that the saturation notion cannot be made to support the heavy burden placed upon it by long-wave exponents.

In our view a critical gap in establishing the recurrence of a long cycle is the absence of a clear economic mechanism that causes the system to move upward from its lower turning point.(34) The most forceful attempt to fill this lacuna is Gerhard Mensch's book, 'Stalemate in Technology'.(35) Mensch asserts that innovations tend to be bunched during depressions, even though they build upon knowledge that was generated at an earlier stage and could have been embodied in innovations at an earlier date.(36)

There are two sources of difficulty with Mensch's case for the bunching of innovations during periods of stagnation and depression: the questionable nature of the data base that he employs for his purpose and the inconclusiveness of his economic argument for the pattern of bunching that he purports to find. On the identification and dating of inventions and innovations, there are still no satisfactory criteria the application of which would command widespread consensus.(37) The issue is, unfortunately, fundamental because, some would argue, it is only certain peculiarities of Mensch's categorization that allow him to say that 'up to now Western industries have largely dwelled on the swell of basic innovations that came in the 1930s, and in the 1950s and 1960s very little basic innovation push developed on which we could expand in the 1970s and 1980s' (pp.30-31). In fact it would not be difficult to compile a most impressive list of major innovations that occurred in the 1950s and 1960s, drawing from such burgeoning fields as pharmaceuticals, electronics, computers, instrumentation, communications, nuclear power, materials (synthetic and otherwise) etc. Regrettably, lists of inventions and innovations, and their appropriate dating, remain notoriously subjective and arbitrary, and cannot yet support the burdensome structure of argument placed upon them. To invoke Mensch's own plaintive note: 'tot homines, quot sententiae'.

The economic argument for the bunching of innovations is also far from convincing. Mensch's view is that innovative activity is finally undertaken in the depths of an economic down-turn because, in their inability to generate profits from older and better-established products whose markets have been saturated, businessmen eventually turn in desperation to new products. Thus, the adverse economic circumstances of the depression are perceived as providing a push into new product lines that were rejected under earlier, more favourable economic circumstances.(38)

It is not, however, easy to see how economic adversity itself would accelerate such substantial long-term financial commitments before a lower turning point had been reached. On the contrary, there is much in the economic logic and in the environmental conditions of a depression, as it confronts the individual members of the business community, that would seem to militate against it. Although perceptions of risk may be expected to become more favourable once it is clear that the recovery phase has already begun, it is difficult to see how depression conditions could provide such an improvement either in perception or opportunity. In fact, depressions tend to make entrepreneurs and managers (as well as sources of financial capital) overly cautious, committing resources which are just adequate for marginal improvements upon the existing technology. This practice excludes, a fortiori, major projects, which require long planning horizons and favourable assessment of risk.

It does not exclude committing resources to the basic design of new products that, in spite of providing the eventual development basis upon which whole new technological trajectories can be built, do not in themselves produce significant macroeconomic effects. Research and development expenditures during phases of retardation (or depression) provide, at most, the technoloigcal basis for sustained growth during expansion. This is so to the extent that such commitments do not necessarily lead to large-scale investments of a kind associated with major innovations. Research and developent expenditures should therefore appropriately be viewed as a necessary but not a sufficient condition for bringing about the commercialization of new products and the actual application of tew techniques of production.(39) Other conditions have to be present, in the form of acceptable perceptions of future risk and returns, before firms will commit large amounts of resources to the construction of new plant capacity.

Of course, the commercialization of some new products involves long gestation periods (e.g. commercial jets or telecommunications systems) and setting up a new facility for the production of even well-established products will often require a similar long gestation period (e.g. a conventional hydroelectric power plant). The decision to proceed on such projects, in anticipation of some future demand, may well necessitate commencement during the depression phase of business cycles. This is, however, very different from the assertion that it is the depression that is responsible for initiating the new undertaking.

In addition, whatever economic cogency Mensch's timing argument might conceivably retain with respect to product innovations, such arguments are of no obvious relevance with respect to process innovations. They can hardly suffer from the 'crowding out' phenomenon that might afflict new products during prosperity phases of the cycle. Indeed, the one phase when the construction of the new plant and equipment embodying process innovations might be least welcome and feasible would be during a depression when, presumably, firms are already suffering from excess productive capacity and facing additional financial constraints. Here again it is difficult to understand the economic logic that would present such innovations as providing the momentum that would generate the lower turning point.

Thus, we reject as economically implausible and unsubstantiated Mensch's view that depression conditions themselves are responsible for the innovations that, in turn, bring the depression to an end. In the specific instance of the 1930s, we would argue that there were numerous other stimuli. The ominous prospects of large-scale warfare in Europe (and eventually war itself) concentrated minds wonderfully upon innovations of great military significance, as in the growing commitment of government funds to the development of the jet engine, radar, and substitutes for especially scarce strategic materials. The ongoing thrust of scientific research in the biological and chemical sciences opened up numerous specific innovative possibilities in fields such as synthetic fibres and pharmaceuticals. The continuing diffusion of an 'old' innovation, the automobile, operated as an increasingly powerful stimulus not only in automobiles directly but in a number of ancillary industries such as glass, rubber, metallurgy and petroleum refining. This bring us, in a sense, full circle to the issue of causality, which we regard as far more complex and multifaceted than made out to be in Mensch's account.

Conclusions

What we have attempted to show is the far-from-trivial requirements that are necessary in order to demonstrate that technological change, in conjunction with macroeconomic factors, can indeed be the preponderant force behind long waves. Having made these conditions explicit, we have also argued, first, that none of the present-day authors who work within a neo-Schumpeterian paradigm has clearly specified the causal links connecting innovation, investment and aggregate rates of growth; second, that the complexities entailed in the timing of the diffusion process, with its technological and macroeconomic determinants, are such that the requirements they impose upon a technologically induced long-wave theory are very stringent, to say the least; third, that the essential exercise of measuring the impact of a set of major innovations upon the economy as a whole has yet to be undertaken by any long-wave proponent. In spite of its difficulties, such an exercise is an important necessary step to add credibility to the notion of long cycles. Finally, it has yet to be shown why the factors responsible for a Kondratiev and its turning points should be expected to have a recurrent character.

We feel we are now entitled to conclude that the conceptual framework of a model of long waves in economic growth, which has at its core the process of technological innovation, has still not been adequately formulated. If long waves are to become a credible notion and serve as a useful analytic framework to understand changes in capitalism over time, there is a clear need to specify their conceptual underpinnings in a more cogent and precise way, in particular the theoretical adequacy of the idea that large or widely diffusable innovations (together with the related investment flows) are responsible for the generation of Kondratievs. Until such a model is developed, the assessment of its historical validity remains unresolved.

We close, therefore, on a sceptical note or, at the very least, on a verdict of 'unproven'. At the same time, we trust that the route that we have travelled in arriving at our present position has been one that has enlarged awareness of the extreme complexity of the connections that link technological innovation, structural change and the long-term dynamics of advanced capitalist economies.

Notes

1. It is interesting to note, however, that poor economic performance in the past decade or so has not been confined to capitalist economies, although it has certainly been pervasive among them. Indeed, the socialist bloc countries of eastern Europe have shared the abysmal economic performance of capitalist countries.
2. Among those who expressed some belief in the existence of long waves are W.S. Jevons, 'Investigations in Currency and Finance', London, Macmillan, 1884; K.

Wicksell, 'Geldzins und Guterpreise', Jena, Fischer, 1898 (an English translation appears as 'Interest and Prices: A Study of the Causes Regulating the Value of Money', New York, Augustus Kelley, 1965); G. Cassel, 'Theoretische Sozialokonomie', Leipzig, C.F. Winter, 1918 (an English translation appears as 'The Theory of the Social Economy', New York, Harcourt Brace, 1932); J. van Gelderen, 'Springvloed - Beschovwingen over industrieele ontwikkeling en prijsbewegigng', Die Nieuwe Tijd, April, May and June 1913, 18, pp.254-77, 370-84, 446-64; and S. de Wolff, Prosperitats und Depressionperioden, in O. Jensen (ed.), 'Der lebendige Marxismus: Festgabe zum 70 Geburstage von Karl Kautsky', Jena, Thuringer, 1924, pp.13-24. For an excellent annotated bibliography on long waves, see K. Barr, Long Waves: A Selective Annotated Bibliography, 'Review', Spring 1979, 4, pp.675-718. (1979).
3. N. Kondratiev, The Major Economic Cycles, 'Voprosy Koniunktury', 1, 1925.
4. N. Kondratiev, The Long Waves in Economic Life (a complete translation of The Major Economic Cycles), 'Review', Spring 1979, 4, p.520 .
5. As Kondratiev noted, when discussing causality of such long-term movements, 'the material basis of the long cycles is the wearing out, replacement and expansion of fixed capital goods which require a long period of time and enormous expenditures to produce. The replacement and expansion of these goods does not proceed smoothly, but in spurts, another expression of which are the long waves of the conjuncture... The period of increased production of these capital goods corresponds to the upswing...conversely, the slowing down of this process causes a movement of economic elements toward the equilibrium level and below it. It must be stressed that the equilibrium level itself changes, in the process of cyclical fluctuations, and shifts, as a rule, to a higher level' N. Kondratiev and D. Oparin 'Major Economic Cycles, Moscow, Krasnaia Presnia, 1928, pp.60-61), as quoted by R. Day, The Theory of the Long Cycle: Kondratiev, Trotsky, Mandel, 'New Left Review', Sept.-Oct. 1976, 99, p.76; and G. Garvy, Kondratieff's Theory of Long Cycles, 'Review of Economics and Statistics', November 1943, 25, p.208.
6. P. David and P. Solar (A bicentenary contribution to the history of the cost of living in America, in P. Uselding (ed.) 'Research in Economic History', vol.II, Greenwich, Connecticut, JAI Press, 1977), in their careful review of the changes of the cost of living in the USA for the 200 year period (1774-1974) do, in fact, identify long cycles in the rate of change of consumer prices lying in the range between 36-60 years. However, they find no counterpart in real phenomena for these long cycles which they attribute to wars and gold discoveries. Similarly C. van Ewijk (A spectral analysis of the Kondratieff cycle, 'Kyklos', 35 (3) 1982), after employing a spectral analytic test of price and volume series for Great Britain, France, Germany and the United States, confirms only the existence of long cycles in prices, but not in output. An earlier work of van Ewijk (The long wave - a real phenomenon?, 'De Economist', 129, 3, 1933) also presents a reasonably strong case against long waves as a phenomenon in 'real' variables. For the most carefully articulated empirical case for the existence of long waves in output, see J.J. van Duijn 'The Long Wave in Economic Life', London, George Allen & Unwin, 1983).
7. See, for example, C. Freeman et al., 'Unemployment and Technical Innovation', London , Frances Pinter, 1982, which is among the best of the contemporaneous statements on long waves; J.J. van Duijn, 'Die Lange Golf in de Economie', Asen, Van Goscum, 1979, and Fluctuations in Innovations over Time, 'Futures', August 1981, 13, pp.264-75; and G. Mensch, 'Stalemate in Technology: Innovations Overcome the Depression', Cambridge, Ballinger, 1979.
8. 1979, op.cit., p.543.
9. Kondratiev's main academic critic was D.I. Oparin, yet it was Trotsky who set the tone for much of the Russian debate on the 'prospects of world economy'. Trotsky asserts as early as 1923 that 'for those long (50 year) intervals of the capitalist curve, which professor Kondratiev hastily proposes also to call 'cycles', their character and duration is determined not by the internal play of capitalist forces, but by the external conditions in which capitalist development occurs. The absorption by

capitalism of new continents and countries, the discovery of new natural resources, and, in addition, significant factors of a 'superstructural' order, such as wars and revolutions, determine the character and alteration of expansive, stagnative or declining epochs in capitalist development' (cited by R. Day, op.cit., p.71). Implicit here is the idea that distinct growth-generating factors are associated with different phases of capitalism. Thus the long waves, if observable, would either be driven by extra-economic phenomena, or by unsystematic (or non-recurrent) economic impulses. This conception seems to have been at the root of much of the criticism directed at Kondratiev by his Russian colleagues, who shared the belief that the Marxist notion of evolution precluded the continual reproduction of a mode of production nwithout its qualitative transformation over time. The gist of their argument can be summarized by stating that the qualitative dimension dominates the quantitative one, a proposition which would entail looking at waves in their singularity, which would be clearly at odds with the idea of a long wave pattern.

10. It may be added that, in spite of substantial differences among them, some present-day advocates of long cycles - W.W. Rostow (Kondratieff, Schumpeter and Kuznets: trend periods revisited, 'Journal of Economic History', vol.35, December 1975; 'The World Economy: History and Prospect', Austin, Univ. of Texas Press, 1978), E. Mandel 'Late Capitalism', London, New Left Books, 1975; Explaining long waves of capitalist development, 'Futures', vol.13, August 1981) and J. Forrester (Growth Cycles, 'De Economist', vol.125, no.4, 1977; Innovation and economic change, 'Futures', vol.13, August 1981) - seem to share the Kondratiev view that innovations are, somehow, disciplined and structured, and have their timing determined by, such long-term movements. As Forrester has put it: 'I believe that the long wave strongly influences the climate for innovation...but I do not see innovation as causing the long wave...instead I see the long wave as compressing technological change into certain time intervals and as altering the opportunities for innovation ' (1981, op.cit., p.338.) Similarly, Mandel also denies that 'innovations create more or less automatically an expansionary long wave' (1981, op.cit., p.334).

Although regularly using terms like 'Kondratievs' and 'Long Cycles', Rostow is not primarily concerned with aggregates or overall movements. His focus is consistently sectoral, especially the changing sectoral composition of investment. His main emphasis is sectoral shifts in relative prices (terms of trade) between agriculture and raw materials, on the one hand, and industrial output on the other, which have assumed a long wave pattern. These relative price shifts have, of course, influenced profit rates by sector and, in turn, have accounted for changing patterns in the allocation of capital. In addition, Rostow rejects the emphasis on innovation as the main determinant of overall and sectoral price swings. Note in this connection that W.A. Lewis 'Growth and Fluctuations, 1870-1913', London, George Allen & Unwin, 1978) is more restrictive in his argument for the existence of price swings, insofar as he is looking only at the period 1870-1913. Further, in Lewis (as opposed to Rostow), the Kondratiev swing was primarily in absolute price levels, which was 'accompanied by' changes in the terms of trade between agriculture and industry (p.27).

11. Of course, Schumpeter's definition of innovation was much broader than the mere technological component, but this issue is not explored here for the reason that his long wave hypothesis is grounded upon the technological aspect of the innovation process. We also ignore the considerable shift in emphasis that appeared in Schumpeter's 'Capitalism, Socialism and Democracy', (New York, Harper & Row, 1942), which tended to see technological innovation itself as much more endogenous than the view expressed in his 'Business Cycles' (New York, McGraw-Hill, 1939).

12. The extent to which innovation can be reduced to rational calculation, based upon observable market conditions, is of course a complex question. Schumpeter was fond of emphasizing the social leadership role of the entrepreneur, especially his willingness to take bold leaps into the unknown, and to undertake commitments that cannot possibly be subjected to the ordinary calculus of business decision-making. Thus, for Schumpeter, America's vast western railroad construction meant

'building ahead of demand in the boldest acceptance of the phrase and everyone understood them to mean that. Operating deficits for a period which it was impossible to estimate with any accuracy were part of the data of the problem ' (J.A. Schumpeter, 'Business Cycles', vol.I, p.328.) This view has been challenged by Albert Fishlow, who argued that the pattern of railroad expansion into the midwest during the antebellum period is best understood in terms of a response to existing and not long-deferred and uncertain profit opportunities - a view that is borne out by the fact that railway enterprises earned profits from the start which were fully comparable to profits earned elsewhere in the economy. Fishlow carefully examines the pattern and sequence of road-building on a state by state basis, and demonstrates that they were profitable from the beginning. See Albert Fishlow, 'American Railroads and the Transformation of the Antebellum Economy', Harvard University Press, 1965, ch.4.

13. Note that in Schumpeter's complex treatment of Kondratievs, the links between innovation clusters and the phases of the cycle are mediated by the varying degrees of 'roundaboutness' of the methods of production (time being essential to build new plant and equipment), and by gestation and absorption lags involved in the introduction of innovations into the economic system. Further, Schumpeter's schema implies that output of consumer goods increases most in recessions and revivals whereas output of producer goods increases most in revivals and prosperity, so that total output increases through all the phases of the cycle. In contrast, the price level rises in the prosperity phase, and it falls in the recession phase.
14. Op.cit.
15. Ibid.
16. Ibid., p.64.
17. Ibid.
18. Ibid., p.65.
19. Partly for this reason, too much importance is commonly attached to specifying the year in which an invention may be said to have occurred. Some of these issues are treated in Nathan Rosenberg, Factors Affecting the Diffusion of Technology, ch.11 of 'Perspectives on Technology', Cambridge University Press, 1976.
20. It is interesting to note that, although Schumpeter popularized the notion of long cycles, or Kondratievs, in his three-cycle schema, he was nevertheless careful to assert that his own innovation theory of business cycles did not require cycles of the length postulated by Kondratiev. 'It cannot be emphasized too strongly that the three-cycle schema does not follow from our model - although multiplicity of cycles does - and that approval of it or objection to it does not add to or detract from the value or otherwise of our fundamental idea, which would work equally well or ill with many other schemata of this kind ' (Schumpeter, 1939, pp.169-70).
21. In this discussion we are examining the case of a closed economy and we are therefore ignoring the additional complexities that would arise from the international transmission of long cycles.
22. See N. Rosenberg, On Technological Expectations, 'Economic Journal', Sept. 1976, reprinted as ch.5 of 'Inside the Black Box: Technology and Economics', Cambridge University Press, 1982. For an analysis of the impact of technological uncertainties upon investment activity, see Karl Moene, Timing of Indivisible and Irreversible Projects: Micro Considerations Related to the Instability of Investment Activity, 'Research Papers in Economics of Factor Markets, 38', Stanford University, Nov. 1982, pp.38-43.
23. It does involve recognising that, if we insist upon employing biological analogies, such as life cycle models, in economics, we need to recognise the possibility of rejuvenation of mature or senescent industries - whatever the plausibility of such a notion in the biological realm.
24. The 'bandwagon effect', according to Schumpeter, is observed 'because first some, and then most, firms follow in the wake of a successful innovation' (1939, p.100).
25. On this last point, see Freeman et al., op.cit., ch.4.
26. S. Kuznets, Innovations and Adjustments in Economic Growth, 'Swedish Journal of

Economics, September 1972, pp.437-8.
27. J. Schmookler, 'Inventions and Economic Growth', Harvard University Press, 1966, passim.
28. For the sake of completeness, mention should also be made of the so-called 'lateral effects' arising out of major innovations and their corresponding leading sectors. These effects, not detailed here, would take the form of 'urban overhead capital; institutions of banking and commerce; and the construction and service industries required to meet the needs of those who manned the new industrial structure' (see W.W. Rostow (ed.) 'The Economics of Take-Off into Sustained Growth', London, Macmillan, 1963, pp.6-7).
29. See, in this respect, Fishlow (op.cit.), which is a rigorous and imaginative attempt to measure the economy-wide impact of a single innovation.
30. The economic impact of Schumpeterian 'epochal' innovations would presumably be dependent upon their spatial implications, with sizable population shifts triggered by such innovations being accompanied by large-scale construction and other subsidiary activities. It would still need to be demonstrated, of course, why such a process added up to cycles of the periodicity postulated by Kondratiev, rather than, for example, the much shorter ones postulated by Kuznets for the pre-World War I period.
31. Input-output analysis, however, may be extremely useful in estimating some of the relevant magnitudes. The following discussion draws in part from Rosenberg (1982, op.cit., ch.3).
32. O. Lange, in his review of Schumpeter's 'Business Cycles' 'Review of Economics and Statistics, Nov. 1941, 23, pp.190-93), although readily agreeing that recurrence was a theoretical possibility, argued that 'there is serious doubt whether the Kondratievs can properly be called cycles. Prof. Schumpeter's explanation in terms of the three great waves of innovation in the history of capitalism seems quite correct. But these three waves of innovation appear to be more of the nature of historical 'accidents' due to discoveries in technology than regular fluctuations in the risk of failure... Schumpeter has extended the theory of business cycles, worked out originally with reference to Juglars, rather mechanically to Kondratievs and Kitchins' (p.192). See also S. Kuznets (Schumpeter's business cycles, 'The American Economic Review', vol.30, June 1940) for a comprehensive critique of Schumpeter's three-cycle model as presented in 'Business Cycles', and his sceptical comments on recurrent 50-year cycles.
33. Let us make this point more precise. If we represented the structure of our model economy, and its movement through time, by a system of linear difference (or differential) equations, then a persistent cyclical behaviour would be obtained if one of the following is true: (a) the solution of the system, the complex conjugate roots, has unitary modulus, and the structural parameters are of the right magnitudes so as to generate cycles of appropriate periodicity and amplitude; (b) the modulus of the complex roots is less than unity (the motion of the system being strongly damped), yet random disturbances (not necessarily serially correlated) generate, or are smoothed into, a cyclical pattern; (c) the modulus of the complex is greater than unity (in which case we observe an increasing amplitude over time) or the value of the dominant positive real root is greater than unity, yet the existence of floors and ceilings constrains the movement of the economy in a cyclical fashion. Insofar as the characteristic roots depend on all structural parameters of the system, it would therefore take its complete specification to ascertain what sort of periodic motion it is able to generate, if any. We have not fully specified the model in this manner, not only because of the complexity of the undertaking, but because of the difficulties of ruling out the possibility of a Kondratiev thorugh this sort of formal exercise. Moreover, further examination of the data might not be conclusive, even if long and reliable production time series existed. In particular, as Sidney Winter has reminded us, if our model economy is represented by a set of non-linear dynamic equations, even simple deterministic systems can mimic a periodic motion for many periods and then shift to other motions without any change of parameters or shocks to the

system. Thus, historical evidence that would appear to suggest a particular cycle in the data needs to be interpreted with extreme caution.
34. Kondratiev (1928, op.cit.) did have such a mechanism, however primitive. As we recall (fn.2), the depreciation and ensuing replacement of the most durable and costly forms of fixed capital were at the basis of Kondratiev's explanation of long waves. The capital goods Kondratiev was referring to were 'big plants, important railways, canals, large land improvement projects, etc ', and their investment spurts were connected to the large sums of free loanable capital available at the bottom of the cycle. Thus the lower turning point would come as interest rate levels were driven down by the accumulation of funds by those on fixed incomes, as the general price level declined. The upper turning point occurred with the depletion of these loanable funds and the ensuing rise in interest rates, which would lead to a curtailment of investment.
35. Ballinger Publishing Company, Cambridge, 1979 (originally published in German in 1975 under the title 'Das technologische Patt').
36. Mensch, op.cit., pp.156-7, 193-4. Mensch argues, for example, that 'most of the essential basic inventions later to be applied in the innovative surge of the 1930s were already well known by 1925. Thus we see that the paradox of unused technologies existed even in the 1920s ' (p.156).
37. For most purposes, Mensch makes use of a subset of the inventions listed by Jewkes, Sawers and Stillerman in their book, 'The Sources of Invention' (New York, Norton, 1969). For a careful criticism of Mensch's sources and uses of data, see C. Freeman et al., op.cit., ch.3.
38. 'In the preceding chapter for the 220 years surveyed, the empirical findings showed that surges of technological basic innovations emerged after economies had fallen into a serious crisis and then passed through years of depression. A graphic representation of the fluctuations in innovative implementation shows that there is a damming up of innovative activity until the onset of economic crises, and then innovations break through the floodgates ' (Mensch, op.cit., p.138.)
39. The Charpie Report has been widely quoted for its suggestion that research and development expenditures constitute only a small fraction of the total costs of successful innovations. According to that report a 'typical distribution' might be that research and development activity accounted for 5-10% of the total cost of introducing an innovation, while engineering design accounted for 10-20%, tooling for 40-60%, and manufacturing start-up costs for 10-15%. See Panel on Invention and Innovation, 'Technological Innovation: Its Environment and Management', Washington, DC, Dept. of Commerce, 1967, p.9. Although there is no empirical basis for these numbers and although the actual numbers doubtless vary enormously among firms and among industries, it is sufficient for our purposes to assert that research and development expenditures may be, and often are, only a small proportion of total innovation costs.

2 Structural change and assimilation of new technologies in the economic and social system

Carlota Perez

In this paper we shall first outline a set of hypotheses about the interrelationship between new technologies and economic development. In so doing, we hope to provide at least a partial answer to Nathan Rosenberg's challenge to specify 'the conditions which would need to be fulfilled in order for technological innovation to generate long cycles in economic growth'.(1) We shall then illustrate this general model with the example of the fourth Kondratiev and make some exploration of the possible characteristics of the fourth. Finally, we shall discuss the conclusions which emerge from this analysis, particularly with respect to policy implications.

We start from a somewhat Schumpeterian view of the role of innovation in provoking the cyclical behaviour of the capitalist economy. But, departing at least partially from his view, we postulate that Kondratiev's long waves are not strictly an economic phenomenon, but rather the manifestation, measurable in economic terms, of the harmonious or disharmonious behaviour of the total socio-economic and institutional system (on the national and international levels).

A structural crisis (i.e. the depression in a long wave), as distinct from an economic recession, would be the visible symptom of a breakdown in the complementarity between the dynamics of the economic subsystem and the related dynamics of the socio-institutional framework. It is, at the same time, the painful and conflict-ridden process through which a dynamic harmony is re-established among the different spheres of the total system.

The resulting complementary trends represent what we might call a 'mode of development' understood as a general pattern of growth, based on a set of accepted social and institutional mechanisms, national and international, influencing the operation and evolution of market and other factors.

What provides the direction and shape of the movement are successive technological styles - or, if you prefer, successive quantum jumps in the general best-practice frontier - based on a constellation of interrelated innovations both strictly technical and organizational, the diffusion of which is propelled by the profit motive.

So for us the long waves represent distinct successive modes of development, responding to distinct successive technological styles. However, although we identify modes of development as stretching from trough to trough of each Kondratiev, we propose that technological styles evolve roughly from the peak of one long wave to the peak of the next. This is the crucial point on which medium and long-term forecasting and counter-cyclical policies could be based. We claim that the crisis is brought about by the introduction of a new technological style when - and because - the previous one approaches the limits of its potentialities. Its initial diffusion, up to a certain critical level, both provokes the crisis of the old mode and sets the guidelines for the next mode of development, during the upswing of which the new style will display its full potential.

Kondratiev(2) had certainly mentioned that during the downswing, together with other characteristic phenomena, there was 'an especially large number of important discoveries and inventions in the technique of production ... which, however, are usually applied on a large scale only at the beginning of the next long upswing'. However, Kondratiev emphasized that this and other recurring relationships did nothing more than further confirm the existence of the long waves. He strongly stressed that he did not 'by any means hold that they contain the explanation'.

Thus, as far as the causation mechanism is concerned, Kondratiev does not make any explicit commitment to the role of innovation (at least, not in the article we are analysing). He attempts to demonstrate the existence of long waves, he denies the possibility that they may be due to random factors, and advances his opinion that their causes are 'inherent in the essence of the capitalistic economy'.

For Schumpeter, who sets out to build a theory of the causation mechanism, the cyclical behaviour of the capitalistic economy has a single root cause: innovation, which in turn is due to a privileged agent: the entrepreneur.

By innovation he understands in very broad terms a 'new way of doing things', be it technical or organizational or a 'new product'. By entrepreneurship he refers to the specific function of carrying out an innovation as distinguished from the managerial function of carrying out production along established lines (even though in many cases both functions may be fulfilled by the same person).

His position as to the essential character of these features is so strong that he defines capitalism itself as 'that form of private property economy in which innovations are carried out by means of borrowed money, which in general though not by logical necessity, implies credit creation'.(3)

Assume a state of perfect Walrasian equilibrium and take an entrepreneur who borrows money to introduce an innovation. From the moment credit is created until his product actually gets to the market, all his actions contribute to enlarge demand (purchase of equipment and inputs, hiring workers etc.). If we assume, for the sake of argument, a project of considerable size, we can see how this enlargement would propagate in waves through the rest of the economy inducing a prosperity phase. However, when the new product appears, supply suddenly becomes greater than demand. The process of regaining a new Walrasian equilibrium requires the elimination of certain inefficient 'old' products and producers. This is the role of recession and the reason for its onset. But the new equilibrium is reached at a point which, though lower than the previous peak of prosperity, is higher than the point of departure. And that is how progress would occur in the long run as well as the reason why it takes a cyclical form which, far from being pathological, would constitute the very modus operandi of the system.

Of course this skeleton model is only the core from which the total system in all its elements is seen evolving as an infinitely complex composite of many synchronous and asynchronous waves impinging upon one another. Though circumstances external to the economic system may induce waves of their own, the core cyclical movement is caused by the rhythm of innovation. Notably, single innovations are seen as igniting a process of imitation. The very success of the innovation - measured among other things by the 'windfall' entrepreneurial profits - will induce others to follow with equal or similar products or methods. This would explain why innovations come in clusters. It also comes closer to reality by providing the link between the role of a single innovation, which creates only a small disturbance in the equilibrium of the system, and the much more noticeable wave created by a growing cluster.

Consequently, the main difference between short, medium and long cycles lies for Schumpeter in the relative importance and weight of the specific innovation or group of innovations which 'carry' them. The Kondratiev long waves, in particular, would be carried by a series of interrelated inventions. Each of them would consist of 'an 'industrial revolution' and the absorption of its effects'.

Though conscious of the fact that we do great injustice to the wealth of Schumpeter's thinking, we must confine ourselves to this very brief sketch of his main line of argument. No matter how far one goes into the complexities of his theoretical presentation and his analysis of the historical material, the same objection can be made: for him the systemic process unfolds within the economic sphere conceived as a self-regulating organism which provokes its own disturbances (innovations) and absorbs its impacts by constantly striving toward new higher equilibria. As for the rest of society, it suffers and profits from this permanent process of 'creative destruction'; it is slowly and profoundly transformed; it is sometimes an obstacle, at other times a stimulus, but it is mainly an environment. Social conditions and the institutional framework are conditioning and conditioned by economic evolution, but they do not form a total structure with the economic system. They are therefore excluded from the causation mechanism for cyclical behaviour.

This is, in our opinion, the reason why even though Schumpeter's theory is generally associated with the explanation of crises or great depressions; he in fact gives a much better account of the shorter cycles and recessions than of the deeper long-cycle depressions.

Even his language is revealing of this uneasy spot in his model. He speaks of prosperity and recession when referring to the Juglar intermediate cycles, but he uses the term 'abnormal liquidation' for the path from recession toward the trough of a long wave and 'recovery' or 'revival' for the beginning of a long upswing. In fact, his model does not really provide a 'natural' exit from a depression - so much so that, in spite of his strongly inimical attitude toward outside intervention in the self-regulating economic system, he reluctantly admits that 'the case for government action in depression, especially government action of certain types, remains, independently of humanitarian consideration, incomparably stronger than it is in recession'.(4)

Strictly speaking, if the system worked as Schumpeter says, deep depressions would be abnormal phenomena and their historical regularity indeed puzzling. Bypassing the problem by invoking the idea of an 'industrial revolution' is in conflict with the identification of the market as the absorption mechanism. Presumably, though the innovations do come in clusters, they are not absolutely synchronic in their introduction and the market should be able to absorb them gradually, through short or medium wave-like movements.

So Schumpeter does lay the foundations for a theory of the cyclical nature of the capitalist economy but **not** of long waves.

Model of the capitalist system

We propose that the capitalist system be seen as a single very complex structure, the subsystems of which have different rates of change. For the sake of simplicity we can assume two main subsystems: on the one hand technoeconomic and on the other social and institutional, the first having a much faster rate of response than the second. The long waves would be successive phases in the evolution of the total system or, as we have termed them, successive modes of development. The root cause of the dynamics of the system would be the profit motive as generator of innovations in the productive sphere, understood in the broadest sense as a way of increasing productivity and expected profits from new investment.

Each mode of development would be shaped in response to a specific technological style understood as a sort of paradigm for the most efficient organization of production, i.e. the main form and direction along which productivity growth takes place within and across firms, industries and countries. The particular historical form of such a paradigm would evolve out of certain key technological developments, which result in a substantial change in the relative cost structure facing industry and which, at the same time, open a wide range of new opportunities for taking advantage of this particular evolution. In essence we assume a strong feedback interaction between the economic, social and institutional spheres, which generates a dynamic complementarity centred around a technological style as roughly defined above. The upswing of the Kondratiev wave would be sustained and stimulated by the harmonic evolution of such complementarity up to the point where the underlying technological style approaches the limits of its potential for increasing productivity.

To surmount this barrier, through trial and error, a new technological style emerges in the productive sphere to which the prevailing social and institutional framework is no longer suited. The new dynamics introduced in the system produce greater and greater disruption in the previously expected evolution of most markets, gradually transforming the social fabric and rendering the institutional mechanisms - which have a high degree of natural inertia, strengthened by the confidence of previous successes - more and more obsolete and counterproductive. This process would be visible as the downswing of the Kondratiev wave, eventually leading to a crisis of the whole system.

The structural crisis thus brought about is, then, not only a process of 'creative destruction' or 'abnormal liquidation' in the economic sphere, but also in the socio-institutional. In fact, the crisis forces the restructuring of the socio-institutional framework with innovations along lines that are complementary to the newly attained technological style or best-practice frontier. The final form the structure will take, from the wide range of the possible, and the time span within which the transformation is effected to permit a new expansionary phase will, however, ultimately depend upon the interests, actions, lucidity and relative strength of the social forces at play.

This rough summary of the form in which we see the evolution of the system should serve to suggest the way in which we envisage prediction to be possible at a time when the usual extrapolations seem powerless. If the characteristics of the new technological style (which is already in place) can be identified and the trends created by its diffusion - both in the economic and extra-economic spheres - can be disentangled from those belonging to the waning style, then the general lines of transformation could be prefigured and serve as criteria for purposive action. We shall return to this point in the concluding section of this paper. Let us now proceed to give a more precise definition of the elements of the model we have sketched.

Model elements

Technological styles

We have been speaking of 'technological styles'. Others may prefer to call them 'technoeconomic paradigms' or 'patterns'. It is not easy to find the ideal term with which to convey the features of the phenomenon we are trying to describe. By 'technological style' we mean a sort of 'ideal type' of productive organisation or best technological 'common sense' which develops as a response to what are perceived as the stable dynamics of the relative cost structure for a given period of capitalist development. As long as the expected pattern of evolution of the relative costs of various types of material inputs, various types of equipment and different segments of labour skills follows the expected trends, managers and engineers will apply what becomes the 'technical common sense' to make incremental improvements along the natural trajectories of the technologies in place, or radical technological changes in those branches of production of goods or services which have not yet achieved the 'ideal type' of productive organization.

So for a given period, with a given set of expected trends in the relative cost structure, more and more branches of the economy will tend to apply the prevailing technological style understood as the most rational and efficient way of taking advantage of the general cost structure. The establishment of such a style or paradigm is grounded on the introduction of a cluster or constellation of interrelated innovations both technical and managerial which lead to the attainment of a general level of total factor or physical productivity clearly superior to what was 'normal' with the previous technological style.

This quantum jump in productivity can be seen as a technological revolution, which is made possible by the appearance in the general cost structure of a particular input that we could call the 'key factor', fulfilling the following conditions:

a) clearly perceived low - and descending - relative cost;
b) unlimited supply for all practical purposes;
c) potential all-pervasiveness;
d) a capacity to reduce the costs of capital, labour and products as well as to change them qualitatively.

The conjunction of all these characteristics in a particular type of input, which, from a technical point of view, was probably available long before, occurs as a response to a persistent demand for technologies capable of surmounting the limits of the technological trajectories based on the use of the prevailing (or previous) 'key factor'.(5) However, once this conjunction of characteristics crystallizes and the evolution of the relative cost structure is modified in a manner generally perceived as long-term, engineering and investment behaviour tend to shift towards new technological paths. We then witness not only the establishment of a new 'best productive common sense' which strives to get maximum advantage of the new key factor, across wide families of related or apparently unrelated technologies, but also a sustained bias in favour of its intensive use, both in radical and in subsequent incremental innovations.

It is important to note that, as mentioned above, the appearance of the new 'key factor' and the technological style that takes shape around its characteristics are phenomena that occur near the peak and during the downswing of the previous Kondratiev and that the transformations they generate in the productive sphere through their gradual diffusion will demand complementary innovations in the social and institutional spheres in order to give way to a new long-wave upswing.

We suggest the role of key factor was played by low-cost coal and steam-powered transportation in the second Kondratiev; by low- cost steel for the third; low-cost energy, in the form of oil and energy-intensive materials for the fourth; and is now being played by low-cost microelectronics on the way towards the fifth upswing.

As examples of what constitutes a technological style one might turn to the most recent and best known, which would be those shaped by low-cost energy between the third and fourth Kondratievs and, as we propose, by low-cost microelectronics between the fourth and fifth. The first would be the extension of the continuous flow concept of the chemical industry to the mass production of discrete identical units made with energy-intensive materials (the prototype of which was Henry Ford's assembly line), complemented on the organisational level by a sharp separation of management and administration from production along the general pattern of Taylor's ideas of scientific management. The second, taking advantage of the characteristics of microelectronics, could perhaps be the flexible batch production network where all activities (managerial, administrative, productive, etc.) are integrated in a total information-intensive system to turn out information-intensive products.

Carrier, motive and induced branches

This brings us to another element of the model we are sketching, namely the contention that the emergence of a new technological style is accompanied by a general shift in investment patterns from the areas that were best adapted to the old style towards those most amenable to the new paradigm. This shift would result in a change in the relative importance of the different branches and in the specific intersectoral relationships. In concrete terms we think that it is possible to distinguish for each technological style - and therefore for each Kondratiev upswing - a specific network of interbranch relationships which describes the main characteristics of the distribution of production between large and small firms and in relation to their weight in the gross product.

Essentially there would be three main types of branches determining the shape and rhythm of economic growth for the period:

1. The **carrier branches**, which are those that make intensive use of the key factor, are the best adapted to the 'ideal' organisation of production, induce a great variety of investment opportunities up and downstream (among them, and most important, great infrastructural investment of specific sorts), and therefore become the vectors of the technological style, having great influence in the general rhythm of economic growth
2. The **motive branches**, which are responsible for the production of the key factors and other inputs directly associated with them, and therefore have the role of maintaining and deepening their relative cost advantage. Thus, while the motive branches create the conditions for the development of the technological style, the growth of their own market depends on the rhythm of generalisation of the style across industries.
3. The **induced branches**, which are both a consequence of and complementary to the growth of the carrier branches (and often use precisely the types of labour displaced by these), only reach their full flourishing and multiply in bandwagon fashion once the necessary social and institutional innovations have opened the way for the upswing and the generalization of the new technological style.

Of course, there are many other branches which produce necessary goods under older, less productive, technological styles or with 'odd' highly specific technologies which are not (or not yet) generalizable. Some of the first are able to get on the bandwagon of the prevailing style through technological innovations, and the general tendency is to try to achieve as much as possible in that direction. But the main argument is that the complementary growth of the carrier and motive branches is the engine that moves the economy and that those branches will tend to be increasingly concentrated in the hands of the largest firms for the period.

Upswing characteristics

To summarize, we suggest that the upswing of a Kondratiev long wave begins when a harmonic complementarity has been achieved through adequate social and institutional innovations, between the 'technoeconomic paradigm' which emerged and developed in the previous Kondratiev peak and downswing, and the socio- institutional climate. This unleashes the swarming process and generates the wave of infrastructural investment that induces the attainment of full growth potential through accelerated diffusion and ultimate generalization of the paradigm.

It is a period of bandwagon effects when, one after another, all productive units - and even social activities of all kinds - tend to apply what is then generally considered as the 'optimal or ideal form of productive organisation'. A particular form of growth stabilizes; a particular way of life takes shape for the different segments of the population; a set of international investment, production and trade patterns evolves; utterly refined statistical models of the economy can be made (and can work); economic science can develop with relative confidence with caeteris paribus assumptions; the trajectories of a large cluster of technologies become 'common sense' and seem to belong to the 'nature of things'; state policies (be they laissez faire or Keynesian or whatever) are seen more as objects of refinement than of radical change because their effectiveness seems to have been 'demonstrated'.

Now, if we remember that the technological style that is then in the process of generalization throughout the system had been introduced during the previous wave - shifting investment of large firms into those branches that have now become the 'carrier' and 'motive' branches, or allowing the appearance of new firms that quickly reach high growth - we can reasonably assume that it is in these branches that the first symptoms of exhaustion of the technological trajectories will be felt.

Presumably then, these would be the most likely to start searching within the large universe of the technologically feasible, though perhaps not yet economically profitable, for new products and new processes that are either labour or materials or capital saving, or seem to offer potential growth prospects. Some of these may result in outright fiascos, others would be the early prototypes of a possible future technological style, placing strong emphasis on (and probably high investment in) developing the cost-cutting potential of the future key factor.

Presumably also, since the limits to a particular technological trajectory tend to translate for the firm into a reduction of the rate of profit or - that which is a similar phenomenon - in a decrease in the expected rate of return of further investment along the same lines, the search for new profit opportunities might not be directed at investment in risky technological innovations but rather towards mergers and acquisitions or less orthodox speculative activites in whatever is found suitable in the particular period (from the mid-1960s to early 1970s there were waves of mergers as well as speculation with money and raw materials; there was persistent recourse to refined manipulation such as 'transfer pricing' and 'leads and lags' in international payments, as well as to developments such as tax havens and other non-productive disruptive practices).

Here a brief reference to Mensch's[6] approach to the theory of long waves should be made. For him investments in alternative types of capital goods are made as a result of 'systematic downgrading in operative value' of existing fixed capital in plant and equipment. We fully agree with this contention but part company with him, following Freeman, Clark and Soete,[7] when it comes to the difference between process innovation and product innovation. Whereas Mensch holds that basic innovations are made in the depths of a downturn, we would contend that the main process innovations (together with those associated with the key factor and the main new organizational paradigm) could well have been made during the later part of the upswing and the beginning of the downswing.

So by the time depression arrives, the new generation of equipment and the organizational pattern that accompanies it is already in the market, and what you get is the application of this equipment mainly to product innovations. It is then vital to distinguish between the initial diffusion of a technological style which is made with 'idle' capital in a period of prosperity (and can be as primitive and costly and risky as Rosenberg[8] suggests) and the further diffusion of a tested technological style which is the most natural investment choice in a period of depression, if and when new investment is to be made in those conditions.

So our contention is that once the initial successful crystallization of the main elements of the new set of technologies has taken place, the peak of the Kondratiev is produced, as the conjunction of the attainment of the old best productive frontier by most of the economy (including the laggards) and a certain degree of diffusion of the new paradigm - within the old mould.

Thus, the peak of the long cycle appears as a kind of economic frenzy of a relatively short duration, but appearing as the promise of everlasting upward progress while the old branches are still joining the bandwagon and new products and processes associated with the emerging technological style produce one big success story after

Structural change and assimilation of new technologies

another. This situation creates unwarranted expectations as to the health of the system and its unlimited opportunities, and it also tends to give undue confidence in the institutional mechanisms, reinforcing their rigidity and inertia. It is in the midst of this high growth situation that the seeds of the contraction are sown.

Downswing characteristics

The descent of the Kondratiev wave sees the exhaustion of the new product and process investment opportunities associated with the prevailing technological style at the same time as the exhaustion of the technological trajectory of the carrier branches (even as their output may continue to grow with inflationary trends). Thus, there is a reduction in the capacity of the motive branches to continue maintaining the relative cost advantage of the key factors, not only for similar technical reasons but also because their main sources of market growth are contracting.

At the same time, those segments of business whose growth potential had seemed unhampered and those of labour whose job and earnings prospects had been more or less 'guaranteed' during the upswing are the hardest hit (a shift which might, by the way, hint at why they tend to support 'strong' solutions to return to 'order').

As the various disequilibria manifest themselves in the various markets (labour, inputs, money, equipment) as a result both of the contraction in the old dynamics and of the unexpected market trends generated by the new investment patterns, more and more pressure is put on the state to find new means of stimulating and managing the economy. The Keyneses and the Schumpeters offer radically new theories and the Roosevelts and the Hitlers establish radically new economic and political management mechanisms, while many others just offer to apply sternly more of the same old successful recipes.

The downswing is then a period of experimentation at all organizational levels of society, characterized by the proliferation of reassessments, proposed solutions and trial-and- error behaviour stimulated by the increasing gravity of the crisis. All this in the face of the weight of tradition, of established ideas, of vested interests and other inertial forces which actively oppose the required transformations.

For the working population it is generally a period of great suffering, as it is they, together with the weaker countries on the international level, who tend to carry the burden of the reaccommodation of the system. (We shall refer later to the shift in occupational profile and in international relative cost advantages which accompany each paradigm shift.)

Meanwhile, on the economic level, the firms that are able (relatively) to escape the crisis are those related to the production or use of the new key factor, which becomes more and more visible in the relative cost structure. It is towards these areas that new investment tends to go, intensifying the disruptive effects of the new technological style and sending signals in all directions for the adequate social and institutional changes required.

The fourth Kondratiev upswing

We shall now attempt to illustrate this general outline with the specific example of the 'assembly line' technological style which we propose shaped the fourth Kondratiev upswing.

We have defined a technological style as a kind of paradigm for the most efficient organisation of production, and have also held that this generates a particular pattern of interbranch relationships also related to the distribution of production between large and smaller capital. We now want to add that each technological style generates a typical pattern of transformation in the occupational structure, and a set of distinct trends in the spatial distribution of production on the international and national scales.

In what follows we shall concentrate on the way the occupational structure is affected by the paradigm shift and its diffusion. The chain of relationships important to our purposes is represented in Figure 1, and we must first briefly discuss what is meant by each of the elements and relationships indicated.

The profit motive should encounter no difficulty when presented as the propelling force and the organizing principle of the capitalist system. In this particular chain of events, however, it has been singled out as the criterion for choosing a particular type of equipment and a specific form of organization of production, taking into account the existing pattern of available technology, opportunities, and especially of relative factor costs, including the evolution of the key factor we defined before and of the various skill segments of the labour force.

While the profit motive is the propeller, the technological style is the steering mechanism. In the chain of relationships now under analysis, we focus on its optimal paradigm for the usage of labour both in quantity and quality, i.e. on the 'ideal skill mix' in relation to the total mass of wages and salaries.

Taylorism

In order to illustrate what we mean, let us look at the historical moment at which the seeds of the recently prevailing paradigm were sown: when Frederick Winslow Taylor transformed the productive organization at the Bethlehem Steel yards at the turn of the century. We could consider this event as the invention and first introduction of a social and institutional innovation within the productive sphere.

According to Taylor's description,(9) over 500 labourers worked at

```
┌─────────────────────┐
│    Profit motive    │
└──────────┬──────────┘
           │
           ▼
┌─────────────────────┐
│ Technological style │
│ (evolution in quantity │◄──────┐
│ and quality of labour) │       │
└──────────┬──────────┘          │
           │                     │
           ▼                     │
┌─────────────────────┐   ┌─────────────────────┐
│ Pattern of evolution │   │ Pattern of evolution │
│ of the occupational  │   │ of consumers'        │
│ structure            │   │ product demand       │
└──────────┬──────────┘   └──────────▲──────────┘
           │                         │
           ▼                         │
┌─────────────────────┐              │
│ Pattern of evolution │─────────────┘
│ of income            │
│ distribution         │
└─────────────────────┘
```

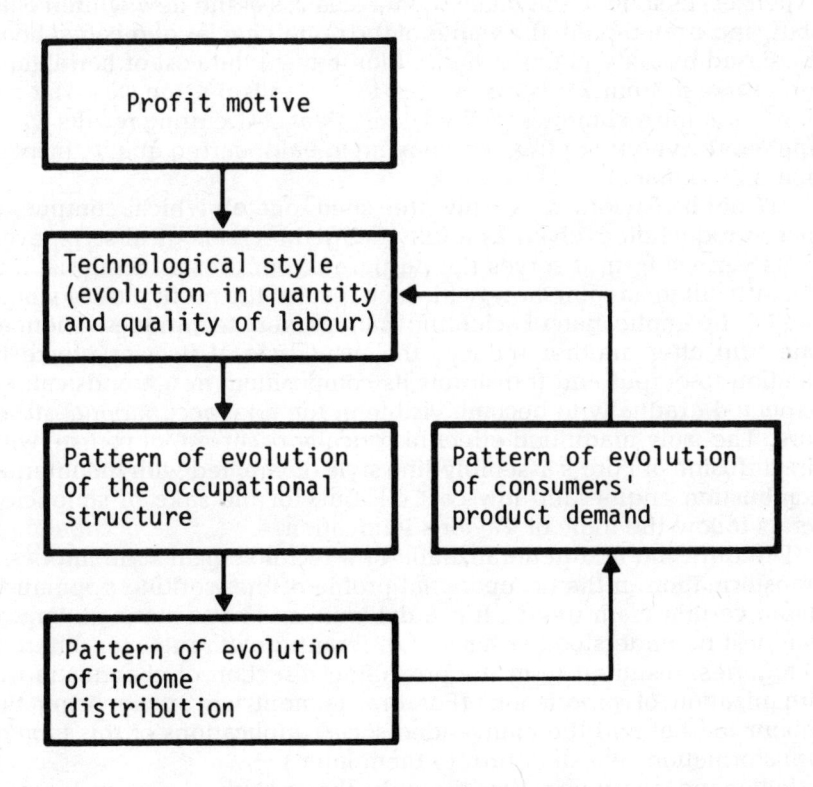

Figure 1. Chain of relationships

different tasks in the yard, in gangs of anywhere up to about 75 men, each group under a foreman. Management merely indicated to the foremen what was to be done and trusted the experience of both foremen and workers (who often used their own tools) to do the job their own way. Three years later there were only 140 labourers in the yard, each accomplishing the work previously done by three or four men but now with standardized, carefully designed company tools, and following strict, standardized procedures determined by time and motion studies.(10) A planning room carefully prepared the following day's work for each individual worker and coordinated all movements in the yard. It was staffed with engineers, time and motion men, draftsmen, a clerical staff, a telephone and messenger system etc. The single group-foremen had been replaced by a set of functional overseers who coordinated, trained, timed and measured work and in general acted as the agents of the planning department.

The new organization, in spite of the new planning and toolroom expenses, in spite of the much higher salaries of the new white-collar staff, and even though the wages of the remaining labourers had been increased by sixty per cent, more than halved the cost of handling a ton of metal, from 7.2 cents per ton to 3.3 cents per ton.(11) And the new scientific techniques yielded equivalent cost-cutting results when applied to everything from bricklaying to ball-bearing quality inspection and to machine- shop work.

Although Taylorism is only the seed out of which continuous mass-production evolved as a fully fledged technological style, even in its earliest form it serves the purpose of our present analysis. It is not difficult to imagine how, with such results, the profit motive would propel the application of scientific management techniques. Then, as one firm after another reduces the usual size of its work force in relation to output and transforms its composition, new trends can be expected gradually to become visible in the total occupational structure. The truly magnified effect historically occurred, of course, with the diffusion of Ford's assembly-line style, combined with the internal combustion engine and low-cost oil. But, for the sake of simplicity, let us follow the logic of Taylor's innovation.

The diffusion and generalization of a technological style implies a transformation in the occupational profile of the working population along certain main trends. It is a dynamic and not a static pattern. It can best be understood as a set of different growth rates for different categories, resulting from the prevailing direction of changes in the organization of production. (For the moment we are ignoring the labour market and the many-sided social implications of this type of transformation. We shall turn to them later.)

Following our illustration through, the growth of a new layer of white-collar workers between managers and foremen, and the reduction of the number of manual labourers required for a given output, introduce a new pattern of evolution in the occupational structure. Initially, the old trends are not eliminated, for much of the growth in the aggregate takes place along traditional lines. The new trends appear simply as counter-trends curbing and transforming the lines of the old pattern. Yet, each business cycle contributes to filter out the old and strengthen and accelerate the application of the new style and the visibility of its consequences on the occupational structure of employment (and of unemployment!).

But this change in occupational structure is accompanied by corresponding trends in income distribution. In Taylor's first experiment the salary mass was more than halved while the workforce was cut to less than a third. But the main change was in the distribution of the globally reduced labour costs. From a three-layer structure of managers, clerks and foremen, and labourers, translating into about one high and one medium salary for every 70 to 80 low-wage labourers, the new organization implied a complex hierarchy of salaries. The ranks of the middle-income groups began to swell. Just as the salary mass was being cut and redistributed within the enterprises, the income mass in the form of wages and salaries was being redistributed in society at large (and many were getting nothing at all).

Again, the evolution in income distribution translates into changes in the pattern of product demand. In our illustration, the new income distribution makes headway in a market sharply divided into luxury and staple goods (food, clothing and rent). The traditional middle class - the small proprietors and the educated few - consumed from (and often catered for) both markets. Now gradually the staple markets begin relatively to shrink, and a new middle-range demand pattern emerges and tends to grow.

But how does this affect the diffusion of the technological style as indicated in Figure 1? At first, we can assume that the new middle layer of salaried workers grows slowly and joins the ranks of the traditional middle class in market behaviour. After all, people choose what to buy from what is available. But the particular evolution of market potential and market stagnation does not go unnoticed by producers. However faulty the information at hand may be, under the conditions we have been following it is not likely that an entrepreneur would launch a new investment to cater for a dwindling market such as cotton textiles, for instance. Instead, he might apply the new potential for productivity increase to turning a luxury good into one accessible to the growing middle layer, which is what Ford set out to do with his Model-T, and many others after him.

Thus the diffusion of a new form of organization of production, requiring a new skill profile, translates into changes in income distribution. These in turn affect the pattern of demand, signalling to producers the general characteristics of the new types of products which would both cater for the growing markets and be capable of being produced with the new technological style - and the process becomes a gradually accelerating feedback loop. This constant propagation increases the disruptive effects in the downswing and the harmony in the upswing.

The introduction of the assembly line as the optimum extension of 'scientific management' contained the crystallization of a change of paradigm for the manufacturing of discrete products along the continuous-flow concept of the chemical industries. It also implied a change in occupational structure which tended to make blue-collar labour more homogeneous and ultimately led to a restructuring of the skill-based trade unions to the branch-based labour unions, while it created a pattern of growth of white-collar labour in an increasing hierarchy. It eventually led to a transformation in the product structure where mass-produced energy-consuming durable consumer goods, made with energy-intensive materials, would gradually be introduced into the sprawling suburban homes which the automobile itself and the expansion of the road network (together with the increase in middle-range incomes) made possible.

As Landes(12) so well expressed it: 'The motor car industry was beginning to play ... (by the end of the interwar period) a role analogous to that of the railroad in the mid-nineteenth century: it was a huge consumer of semi-finished and finished intermediate products ... and components ... it had an insatiable appetite for fuel and other petroleum products; it required a small army of mechanics and service men to keep it going; and it gave powerful impetus to investments in social and overhead capital (roads, bridges, tunnels). At the same time, it posed new technical problems in metallurgy, organic chemicals, and electrical engineering, eliciting solutions that had important consequences for other industries as well '

Thus, there was an accelerated process of diffusion in the main carrier branch - spurred by and spurring the oil industry, already run by a few giants, and rapidly finding lower-cost sources in Mexico, Venezuela and South-East Asia (although the radical cost-cutting was to be made possible by the 'free-flowing' light oil from the Arabian Peninsula in the 1930s). From a peak average of $2.00 per barrel during the war and postwar period (1915-20), the price of oil decreased to an average of $1.35 in the 1926-30 period, and to $0.83 per barrel in the post-crash years (1931-35). The average price of electricity in the USA, on the other hand, had fallen 41% by 1928 with respect to 1902 (or 31% per cent in constant terms).

Technological style

We suggest that in the 1910s and 1920s the technological style that shaped the fourth Kondratiev mode of development, with its carrier and motive branches and its typical skill profile, was already emerging and diffusing.

It is interesting to note that both massive oil production and assembly-line technology were US-based, and that the fastest rates of growth in electricity production and in radio and car sales took place in the USA. The greater weight of the old style and the divided markets of Europe seemed to inhibit the achievement of the full potential for mass production of identical units inherent in the new style. The USA had all the conditions for proceeding unhampered to become the world centre of the new mode of development.(13)

But to make the transition from a system based on the growth of steel, capital goods, heavy electrical equipment, great engineering works (canals, bridges, dams, tunnels) and heavy chemistry - all geared mainly towards big spenders such as other capitalists or governments - into a mass production system catering for consumers and the massive defence markets, radical demand management and

Structural change and assimilation of new technologies

income redistribution innovations had to be made, of which the directly economic role of the state is perhaps the most important.

The big upswing of the world economy after the second world war was a period in which there was a good 'match' between the requirements of a mass production technological style, based on the almost universal availability of cheap oil, and the social and institutional framework within which this technological style could flourish. But this good 'match' was only achieved after a period of deep depression and social turmoil in the 1930s. During the 1930s, it was by no means clear how to achieve a set of appropriate institutional and social responses. As already indicated, the solutions which were then advocated and applied varied across a wide range from fascism to the New Deal and Communism.(14) It was only after the second world war that gradually a mode of development crystallized in the leading industrial countries, which did create the necessary harmonization of institutional framework with technological style.

Socio-institutional structures

Among the main institutional changes which promoted this good 'match' were, on the national level, the major expansion of the role of the state in economic life. The Keynesian policies which, in one form or another, were adopted by most countries led to various demand management mechanisms, both directly through infrastructural, defence and public service spending, and indirectly through income redistribution by means of taxation, interest rate management and massive government employment. More indirect, but equally essential for demand management, was the elaborate public statistics apparatus, which served both public policy and private investment and market forecasting.

Another important socio-institutional change was the rapid expansion of massive secondary and higher education to meet the demand for white-collar, technical and clerical employees, together with the expansion of the various national forms of public health systems. Both were also great sources of employment and hence of income redistribution.

On the more directly economic level, innovations such as large-scale consumer credit methods, the expansion of publicity, the mass communications industry and various forms of planned obsolescence further increased the means of orienting the use of disposable income into intensive consumption of the various goods proper of the mass production style (and later also of those which became the forerunners of the style to come).

The institutional acceptance of the labour unions as legal representatives of the workers (especially in the carrier branches) both fostered the growth of disposable income and stimulated the application of labour-saving incremental innovations, within old plants or in new investment, along the trajectory of the technological style. In general, the evolution of non-union labour wages tended somewhat to follow the trends set by union labour. Thus, one could consider it as an indirect form of demand management.

At the level of the firm, a new 'ideal type' of organization for giant firms emerged, with horizontal integration and a complex managerial system which allowed reaching optimum plant scale under a much larger optimum firm scale. This was described by Peter Drucker(15) in 'The Concept of the Corporation', based on his study of General Motors. Accompanying this organization was the in-house research and development laboratory, which earlier had only developed as a necessary feature of such science-based industries as chemistry and electricity, but could now serve the controlled and sophisticated forms of competition which came to characterize oligopolies.

This particular development was a crucial element of the 'military-industrial complex' which, following the prototype of the Manhattan Project which produced the A-bomb, brought state, scientific, technological and industrial efforts to focus on pre-defined goals, eliciting a flurry of innovation requirements for new materials and processes which could later spin-off from military to civilian uses, once their primitive, more costly stages of development had been borne by defence contracts.

On the international level, the Bretton Woods Agreement established a solid basis for the regulation of intercountry trade and investment (recognizing the hegemony of the USA in the new arrangement), and the Marshall Plan stimulated general international growth of investment and markets. Decolonization broke the empire-based barriers on investment and trade and allowed the energy- and materials-producing motive branches from different developed countries to establish more flexible competitive arrangements to use the low-cost sources available in the developing countries. At the same time, the massive market growth needs of the technological style provoked an increasing number of 'common market ' type agreements, as well as the 'local subsidiary' response of the carrier branches to the developing countries ' tariff barrier policies. However, most international institutions and especially the UN were more of a facilitating nature than actually 'managing' economic growth, when compared to the main national institutions.

This impressionistic and incomplete list of the social and institutional innovations of the fourth Kondratiev mode of development can also be considered - if our hypothesis about long waves is an accepted approximation of the way the process evolves - as a list of obsolete mechanisms in regard to the effective institutions required to unleash the upswing of the fifth Kondratiev based on microelectronics.

Requirements for the next upswing

According to the foregoing hypotheses, the social sciences would today have the tremendous job of disentangling the new trends generated by the already established technological style, with its family of interrelated technologies and more or less visible trajectories, from what are in fact either waning trends due to the exhaustion of the old paradigm, or temporary responses which will disappear once the transition is effected. In a sense, during the upswing we are dealing with statistical distributions where the 'mode' tends to coincide with the 'mean', but in downswings and crises we deal with bimodal distributions where the 'mean' aggregate means very little. Also, during the upswing qualitative factors can be relatively ignored in quantitative measurement, whereas interdisciplinarity and qualitative case study type research are indispensable during downswings.

In particular, the precise identification of the characteristics of the new paradigm is essential in order to point to the institutional solutions which, at the same time as they open the way for the generalization of the new paradigm, find the appropriate solutions to improve the lot of those who would have been its inevitable victims.

Doubtless the range of possible scenarios is quite wide, as indeed was apparent in the trough of the 1930s. However, the new style seems to have a strong transnational dimension, based on the provision of the unprecedented data-management capabilities and telecommunications infrastructure for the efficient management of giant, complex, flexible transnational conglomerates, optimizing factor use and maximizing long-term profits on a planetary scale. Consequently, national solutions of the kind we grew accustomed to in the waning Kondratiev seem ill-adapted to the new paradigm.

The radical productivity increases possible with the new paradigm in the formerly less-productive craft technologies (such as printing) or in batch-production (with numerical control), as well as the general flexibility introduced by computer-aided design and manufacturing, involve profound changes in the relative importance of the various industries. They allow enormous increases in plant scale (where size of plant is no longer necessarily equivalent to size of market for one product, but relates to a stable or changing family of products). These possible increases can involve the integration into one continuous process of various intermediate products, together with the final flexible output, increasing the scaling-up possibilities.

Of course, a technological trajectory is not predetermined, but only the general range of its possibilities. However, the particular potential we are referring to indicates that national markets are a hindrance to full deployment, and also that unless there is an international income redistribution mechanism, the appropriate growth of markets cannot occur.

Finally, there are three aspects of the new technological style, based on microelectronics, that are worth mentioning:

1. It is a materials-saving technology, so the high cost of energy and materials would stimulate its application. So perhaps OPEC, far from being the culprit in the crisis, might be the prototype for a social organization (analogous to labour unions, serving both to stimulate the new technology along its trajectory and as an income redistribution mechanism).
2. It is an information-intensive technology, so the low cost of microelectronics and the increasing capability of data-processors, if and when accompanied by massive growth in telecommunications networks, would open an ever-increasing range of bandwagon application opportunities for information-intensive services for producers and consumers.
3. The flexible product mix allows the application of rapid obsolescence techniques to capital goods both for the office and for the production of goods and services, so the proliferation of small and medium sized producers in developed and developing countries could not but enhance the market prospects of the technology-generative firms (and of the motive branches).

Perhaps an analogy can be made with the third Kondratiev, when large producers did not cater directly to consumers but, based in low-cost steel, they concentrated their growth in the enormous investment in the civil, mechanical and electrical engineering carrier branches, while at the same time allowing for the massive deconcentration of production with free-standing electric motors.

Whether these are some of the appropriate conclusions or not, we can only speculate. Optimism under present conditions is on very shaky ground, but if our hypothesis seems plausible, then it can only be based on the capacity to innovate boldly in the social and institutional spheres and - we contend - on a planetary scale.

Notes

1. N. Rosenberg and C.R. Frischtak, 'Technological Innovation and Long Waves' (Stanford University, Mimeo, Jan 1983), p.3.
2. N.D. Kondratiev, The Long Waves in Economic Life, 'Review of Economic Statistics', vol.17, Nov 1935, pp.105-15.
3. J.S. Schumpeter, 'Business Cycles: A Theoretical, Historical and Statistical Analysis of the Capitalist Process', vol.I (McGraw-Hill, NY, 1939), p.223.
4. Ibid.
5. The actual combination of 'accidental' and 'purposive' events leading to this particular development will not be discussed here. Suffice it to note that the relative autonomy of science leads to a universe of the technologically feasible which is, at any time, much greater than the economically profitable.

Structural change and assimilation of new technologies 47

6. G. Mensch, C. Continho, and K. Kaasch, Changing Capital Values and the Propensity to Innovate, 'Futures', vol.13, no.4, Aug 1981, p.283.
7. C. Freeman, J. Clark, and L. Soete, 'Unemployment and Technical Innovation: A Study of Long Waves and Economic Development' (Frances Pinter, London, 1982).
8. Rosenberg and Frischtak, op.cit., pp.10-11.
9. F.W. Taylor, 'The Principles of Scientific Management' (Norton, NY, 1967).
10. We might note in passing that 'standardization' was a key concept in th prevailing paradigm of the third Kondratiev within which this very influential organizational innovation was born.
11. Again, we might note that we had suggested low-cost steel as the key factor in the third Kondratiev, so any cost-cutting improvements in the motive branch producing it contributed to the upswing.
12. D.S. Landes, 'The Unbound Prometheus' (Cambridge University Press, 1972), p.442.
13. This heavy early commitment and the consequent full adaptation to that particular technological style might help explain why today a country like Japan finds it easier to embrace the new style and make the institutional changes required than does the USA.
14. As far as the mass production technological style that subtends these widely differing institutional and social arrangements, there are no discernible differences, which serves to illustrate the diverse range of alternatives opening before society at each critical phase.
15. P.F. Drucker, 'The Concept of the Corporation' (Mentor, NY, 1972).

3 Cycles, turning phases and societal structures: historical perspective and current problems

Ger van Roon

'La longue durée historique cerne l'innovation'(1)

My contribution to this conference as an historian was not written looking towards the past, but at the present and immediate future. Whereas a lot of historical research is restricted to the past, contemporary history is that approach which attempts to show the origin and development of present-day problem areas through an historical structural analysis. Contrary to the more traditional approach, it starts at the present and works back to the period when a problem of current interest first manifested itself.(2) Although continuity is certainly not neglected, the predominant interest in this approach is in discontinuity. In order that the differences in the historical situation be shown to full advantage, a phase diagram is generally employed. Contemporary history will always be world and social history.

Since 1971 the world's economy has been in a period of stagnation which had already begun in the sixties. During the seventies, this period of stagnation turned into a fast-spreading depression. Although this depression shows a number of similarities with previous ones, there are also major differences.(3) In my opinion, stagnation did not arise as a result of the first oil crisis in 1973. Moreover, the recent fall in interest rates would seem to be not so much a sign of an incipient economic recovery as a symptom of an economic recession that has not yet led to sufficient reform and restabilization.(4) The present crisis is not only an economic but also a social crisis. The position of the socially weaker groups in society has been seriously undermined, unemployment has risen sharply, and the guiding role of the government has increased disproportionately. There is even the threat of anti-participatory restabilization of society. Economic instability strengthens and aggravates political instability, both nationally and internationally.

Taking the depression of 1873-1896, the West German historian Hans Rosenberg has shown how a prolonged depression can lead to serious social and political tensions.(5) With his study of changes in standards and values, Rosenberg has also given a coherent, penetrating and frightening picture of what should be understood in socio-psychological terms as a 'climate of depression'. His analysis could be valuable for an identification of certain present-day developments.

As a result of the current depression, there has been a revival of interest in recent years in the long-wave theory, although it was treated prior to this in several Dutch books on economics.(6) In Holland this series has been the subject of constant attention for some time, with publications by Van Gelderen,(7) followed by De Wolfe,(8) Harthoorn,(9) R. Mees,(10) Tinbergen,(11) I.J. Brugmans,(12) J.A. de Jonge,(13) H. de Vries, (14) and others. The various authors more or less agree on some aspects of the long wave, particularly on the fact that the long wave is a phenomenon of and is caused within industrial capitalist society, although agricultural historians speak of this phenomenon too in earlier periods. (15) However, for current problems it is only of importance after the start of the industrial revolution. This means that theorization on the causes will have to relate to the features of industrial capitalism. A second point of agreement is that the long wave relates to the economy as a whole, in contrast to the coffee, pork or Kuznets cycles, which occur in one market or one sector.

For the rest, their views differ somewhat. The cyclical or mechanistic and the structural approaches differ fundamentally. In the first instance the long wave is viewed as one of the periods in the economic cycle, which develops to a set pattern of about 50 years.(16) In the latter instance emphasis is placed on the alternation of periods of swift growth and relative stagnation with periodic structural changes.(17) The 'discoverers' may see the long wave as being caused by the whole process of change in the capitalist/industrialist economy, but later authors have touched more than once on more or less independent sectoral causes. To an historian the long-wave approach is attractive for more than one reason. Firstly, it offers the possibility of achieving a more satisfactory periodization of late modern history.(18) Living and working conditions can then be examined more specifically by relating them to fluctuations in the economic trend.(19) Another possibility is the use of the comparative method, either between countries in one and the same period, or between different periods of depression.

In addition, other comparisons are possible, namely that between so-called 'turning phases' in the economic trend. By this is meant the transitional period between the trend movements. Too much attention has been concentrated on these movements alone.(20) The term 'turning point' may be not unknown in economic research of the short term, but little research has been focused on turning points in the medium and long term. The views that exist in this field have been inspired notably by the experiences of the thirties. Haberler may have noted this theme already in his 'statusquestionis', commissioned by the League of Nations,(21) but its importance was also pointed out by Akerman, who provided a compact definition of the problem for

Design, innovation and long cycles

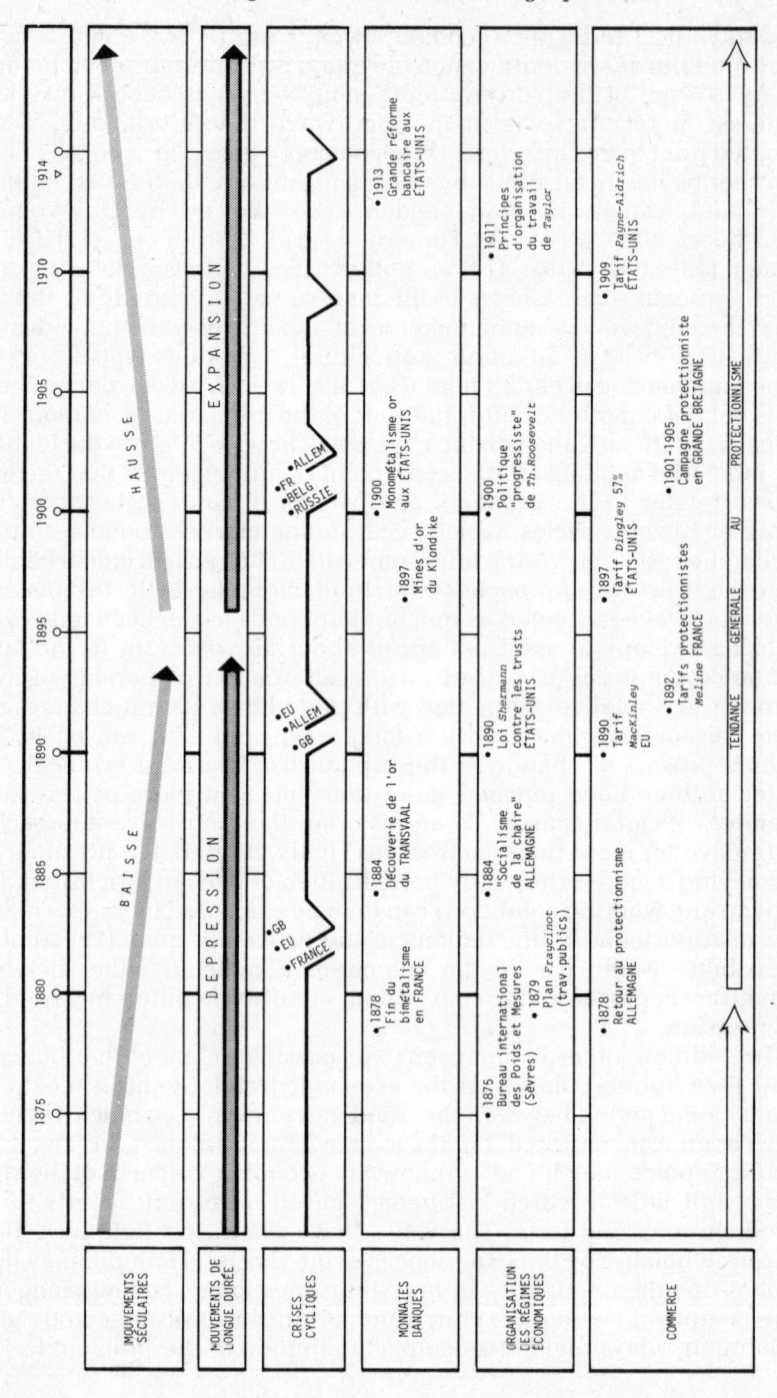

	1875	1880	1885	1890	1895	1900	1905	1910	1914
MOUVEMENTS SÉCULAIRES		BAISSE				HAUSSE			
MOUVEMENTS DE LONGUE DURÉE		DÉPRESSION				EXPANSION			
CRISES CYCLIQUES		• GB • EU • FRANCE		• EU • ALLEM • GB		• FR • BELG • RUSSIE		• ALLEM	
MONNAIES BANQUES	• 1875 Bureau international des Poids et Mesures (Sèvres)	• 1878 Fin du bimétallisme en FRANCE	• 1884 Découverte de l'or du TRANSVAAL		• 1897 Mines d'or du Klondike	• 1900 Monométallisme or aux ETATS-UNIS			• 1913 Grande réforme bancaire aux ETATS-UNIS
ORGANISATION DES RÉGIMES ÉCONOMIQUES		• 1879 Plan Freycinet (trav.publics)	• 1884 "Socialisme de la chaire" ALLEMAGNE	• 1890 Loi Sherman contre les trusts ETATS-UNIS		• 1900 Politique "progressiste" de Th.Roosevelt		• 1911 Principes d'organisation du travail de Taylor	
COMMERCE		• 1878 Retour au protectionnisme ALLEMAGNE		• 1890 Tarif Mackinley EU	• 1892 Tarifs protectionnistes Méline FRANCE	• 1897 Tarif Dingley 57% ETATS-UNIS • 1901-1905 Campagne protectionniste en GRANDE BRETAGNE		• 1909 Tarif Payne-Aldrich ETATS-UNIS	

TENDANCE GÉNÉRALE AU PROTECTIONNISME

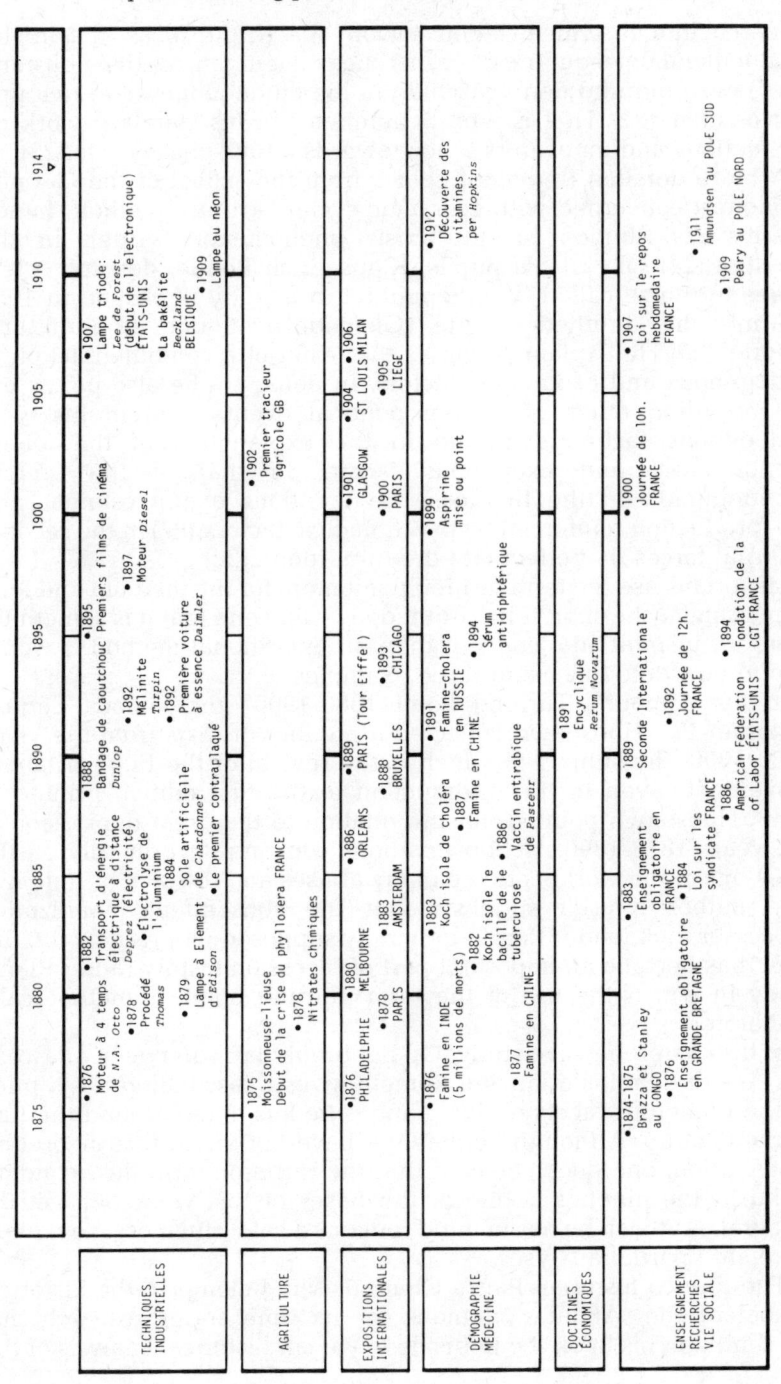

Cycles, turning phases and societal structures

	1875	1880	1885	1890	1895	1900	1905	1910	1914
TECHNIQUES INDUSTRIELLES	•1876 Moteur à 4 temps de *N.A. Otto*	•1882 Transport d'énergie électrique à distance *Deprez* (électricité) •1878 Procédé *Thomas* •1879 Lampe à Elément de *Chardonnet* d'*Edison*	•1884 Sole artificielle •1884 Le premier contreplaqué	•1888 Bandage de caoutchouc *Dunlop*	•1895 Premiers films de cinéma •1892 Mélinite *Turpin* •1892 Première voiture à essence *Daimler*	•1897 Moteur Diesel		•1907 Lampe triode: *Lee de Forest* (début de l'électronique) ÉTATS-UNIS •La bakélite *Beckland* BELGIQUE 1909 Lampe au néon	
AGRICULTURE	•1875 Moissonneuse-lieuse Début de la crise du phylloxera FRANCE •1878 Nitrates chimiques					•1902 Premier tracteur agricole GB			
EXPOSITIONS INTERNATIONALES	•1876 PHILADELPHIE •1878 PARIS	•1880 MELBOURNE •1883 AMSTERDAM	•1886 ORLEANS •1888 BRUXELLES	•1889 PARIS (Tour Eiffel) •1893 CHICAGO		•1900 PARIS	•1901 GLASGOW •1904 ST LOUIS •1906 MILAN •1905 LIEGE		
DÉMOGRAPHIE MÉDECINE	•1876 Famine en INDE (5 millions de morts) •1877 Famine en CHINE	•1883 Koch isole le bacille de la tuberculose •1882 Koch isole le bacille du choléra	•1886 Vaccin antirabique de Pasteur	•1891 Famine-choléra en RUSSIE •1887 Famine en CHINE •1894 Sérum antidiphtérique	•1899 Aspirine mise au point		•1912 Découverte des vitamines par *Hopkins*	
DOCTRINES ÉCONOMIQUES				•1891 Encyclique *Rerum Novarum*					
ENSEIGNEMENT RECHERCHES VIE SOCIALE	•1874-1875 Brazza et Stanley au CONGO •1876 Enseignement obligatoire en GRANDE BRETAGNE	•1883 Enseignement obligatoire en FRANCE •1884 Loi sur les syndicats FRANCE •1886 American Federation of Labor ÉTATS-UNIS	•1889 Seconde Internationale •1892 Journée de 12h. FRANCE •1894 Fondation de la CGT FRANCE		•1900 Journée de 10h. FRANCE	•1907 Loi sur le repos hebdomadaire FRANCE •1911 Amundsen au POLE SUD •1909 Peary au POLE NORD		

Source: P. Dolfaud, C. Gérard, P. Guillaume, J.-A. Lesourd, "Nouvelle Histoire Économique", T.1, pp. 322-323.

research into it, which 'legen sowohl die Krafte bloss, die in der kumulativen Bewegung wirken, als auch diejenigen, die den Stillstand der Bewegung und den Umschlag in die entgegengesetzte richtung hervorrufen'.(22) That is why Schumpeter in his standard work on innovations and innovators tends towards a four-phase cycle.(23)

A more detailed treatment comes from the school of the German economist concerned with economic cycles, Arthur Spiethoff, famed for his contribution on the crisis phenomenon written in the twenties.(24) One of his pupils, Klaus Erich Rohde, devoted a few pages to the so-called 'Wendepunkte' in a study published in 1957 bearing the antithetical title 'Gleichgewicht und Konjunktur-theorie'.(25) He explains them as the result of a combined action of endogenous and exogenous factors. In doing so, he also points out the retarding effect of certain political events. (26) In his view endogenous factors dominate in the explanation of the 'obere Wendepunkt' and exogenous factors in that of the 'untere Wendepunkt'. In the first case he mentions overinvestment and overproduction reinforced by psychological factors;(27) in the second, external forces that effectuate the turnaround.(28)

Historians use the term turning point more for the medium and long term than for the short term. But it does strike one that it is sometimes more an inspirational concept than an operational method. I would like to elucidate this using three examples.

In his 'Labour's Turning Point 1880-1900', the famous English historian E.J. Hobsbawn collected a number of texts from the years 1880-1900, the subject of which is the revival of the English labour movement. Even in the introduction to the first edition printed in 1948, Hobsbawn pointed out the relation to the great depression in the years 1873-1896. His observations concentrate primarily on the reactions of the workers to the many abuses and the lack of state aid. The author himself suggests that the consequent mobilization processes took longer than is generally supposed.(29) As a result, this book has become an important part of recession history rather than a study in which the use of the term 'turning point' is methodically elaborated.

In his comprehensive study 'Op het breukvlak van tween eeuwen', the Amsterdam historian Jan Romein gives a fascinating description of the numerous facets of the changes he has noted around the turn of the century. Although he may not have got round to a theoretical justification, one does perceive that the turnaround in the economic trend in the nineties is one of the bases of this work.(30) Yet the integral approach he has in mind remains a colourful mosaic of almost unrelated partial analyses.

The French historian Pierre Chaunu, who belongs to the historical school of the 'Annales', famous for its time series research, has presented in his book 'Le Refus de la vie' an historical analysis of the present crisis which is not restricted to economic data. He sees the year 1962 as the starting point of the crisis. His observations show that demographic changes played a decisive part in his choice of the year 1962, but he has to admit that this choice was relatively arbitrary.(31)

The term 'turning phase', too, is not unknown in scholarly publications.(32) I understand it to mean a concentration, a clustering of turning points of variables in the medium term. I consider the turning phases as independent periods in the economic trend. Broadly speaking, this gives rise to a four-phase model:

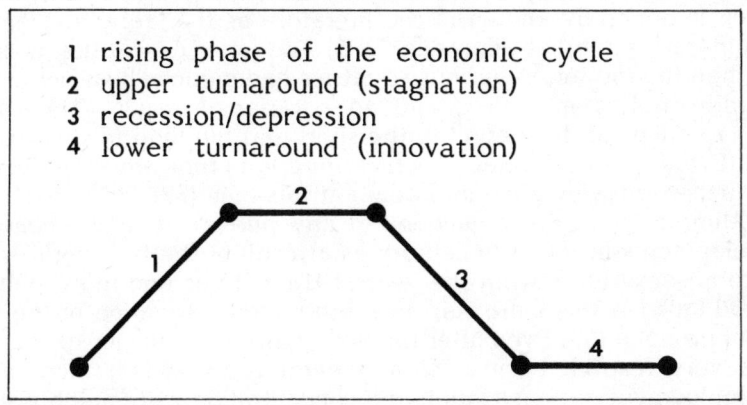

1 rising phase of the economic cycle
2 upper turnaround (stagnation)
3 recession/depression
4 lower turnaround (innovation)

The turning phases form a complex whole of leading and lagging variables. It would be important to know more about the sequence and interrelation of the processes in times of stagnation or innovation. A concrete example may serve to clarify this. Before the economic reverse manifested itself in 1971, there was in several countries about the middle sixties an accelerated drop in the number of births. Whether the effect of the pill should be seen more as a typical example of a late-flowering phase from a period of economic prosperity, or as proof of diminished faith in the future, can be examined in more detail with the aid of analyses of changes in values and standards and saving habits. This example may not only outline what should be understood by a turning phase, but these turning phases may also be expected to fulfil an important structural function. A multidisciplinary, comparative approach would seem an absolute precondition for research in this field.

If one compares the dates of turning points of different authors, big differences are often noticeable.(33) This is understandable, as the same indicators are not always used. If a choice of turning phases is to encounter comparable problems, the basis for this choice could prove stronger through the multitude of indices. Before this is achieved, a start will have to be made with hypothetical and tentative turning phases, whereby differentiations will have to be made per country. To start with, the following periods would seem to qualify for the description 'turning phase':

upper	lower
1869-1879	1893-1903
1911-1921	1936-1949
1963-1973	

The first upper and lower turning phases are the limits of a period which is generally known in the literature as the 'agrarian depression'(34) or the 'great depression'.(35) The latter description is more apt than the former, as various countries experienced a succession of an industrial, a monetary, and an agrarian crisis.(36) The social consequences of this period in the short and the long term are such that divergent developments, which for a long time were ascribed to the first world war, came into being at this time.(37)

Although there are objections to any periodization of economic developments in the 20th century as a result of the two world wars, there is sense in viewing the years 1913-1949 as one big economic period.(38) 1913 saw the start of a protracted stagnation of the nett social product. (39) Even after the first world war one can speak only in a very restricted sense of a general recovery.(40) Stagnation, unemployment, rationalization and concentration were notably striking characteristics of the first half of the 'golden twenties'.(41) On the basis of this approach, a turning period in the years 1926-1936 has been omitted and the years 1911-1921 chosen. The reason for such an early choice for the start of the final upper turning phase was explained earlier.

It would seem advisable to start research into turning phases with an analysis of demographic variables and their turnaround sensitivity in the medium term. After all, demographic fluctuations have to do with economic, social and political as well as socio-psychological factors. An initial investigation(42) showed that birth, marriage and certain migration statistics are major variables for research into turning phases. Data from seven countries were examined: The Netherlands, Belgium, Britain, France, Germany, Sweden and the United States. Turning points were dated at the moment a break became visible or - in fluctuations - where a peak or trough occurred. In the literature it is notably Hatzhold of the IFO Institute in Munich who has drawn attention to the adverse economic effects of a drastic fall in the birth rate over a long period of time.(43) Very recently calls have been heard in the Netherlands for a policy promoting an increase in the population.(44) This 'counterpoint'(45) could perhaps be a pointer to the fact that a turn in the views on this subject is in sight.

Research into the turnaround sensitivity of economic variables in

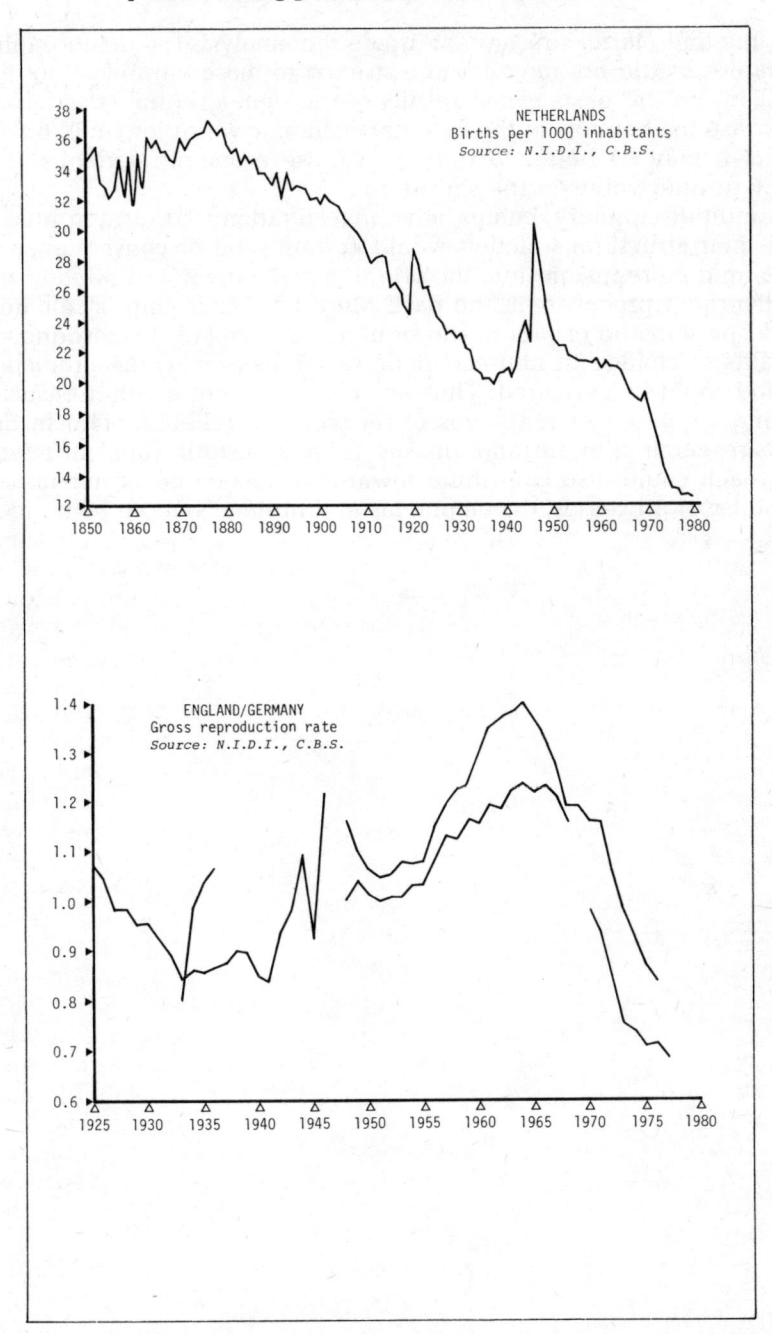

Cycles, turning phases and societal structures

NETHERLANDS
Births per 1000 inhabitants
Source: N.I.D.I., C.B.S.

ENGLAND/GERMANY
Gross reproduction rate
Source: N.I.D.I., C.B.S.

the medium term subsequent upon the analysis of demographic variables would not have to be restricted to those variables that are relevant on the basis of the results of the demographic research. In that case the key role of the external economic variables could be left aside. It may be better to start from those economic variables that show turning points in the short term.

A multidisciplinary, comparative investigation into turning phases and their structural function would certainly be of consequence in obtaining more insight into the origin, development and relationship of divergent processes in the past. More understanding of this field could perhaps be of prognostic significance, too.(46) By recognizing heralds of change in non-economic variables earlier, the prediction period could be extended. Thus one could also get a better basis for testing supposed or real signs of recovery for reliability.(47) In this way research into turning phases from a historicizing, structural approach could also contribute towards a deepening of our understanding of processes, developments and problems of our own times.

UNITED STATES
Immigration
Source: N.I.D.I., C.B.S.

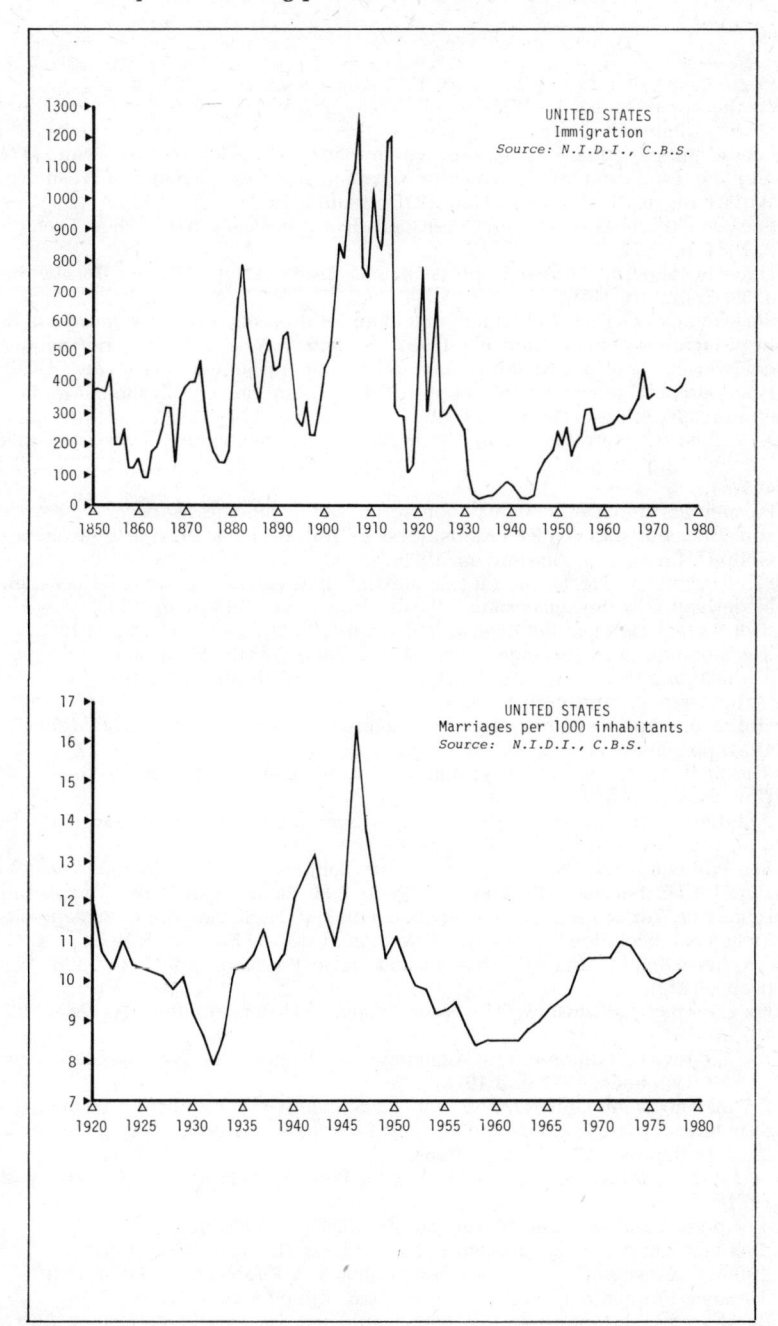

UNITED STATES
Marriages per 1000 inhabitants
Source: N.I.D.I., C.B.S.

Notes

1. Pierre Chaunu, 'Le refus de la vie' (Calmann-Levy, Paris, 1975), p.33.
2. On this approach, see also Geoffrey Barraclough, 'An Introduction to Contemporary History' (Penguin, 1969).
3. For example, Dietmar Petzina, Krisen gestern und heute - die Rezession von 1974/75 und die Erfahrungen der Weltwirt schaftskrise, 'Gesellschaft fur Westfalische Wirtschafts geschichte e.V ', Heft 21 (Dortmund, 1977).
4. See about this last aspect Alan Greenspan, Economic Policy in the 1980s, 'Dialogue', 2, 1981, pp.9-12.
5. Hans Rosenberg, 'Grosse Depression und Bismarckzeit' (Ullstein Taschenbuch 3239, Frankfurt, 1976).
6. For example, G.Th.J. Delfgaauw, Inleiding tot de economische wetenschap, deel II, 'Macro-economie' (Delwel, 1973); S. and F.A.G. Keesing, Het moderne geldwezen, deel I, 'Macro-economische uitgangspunten', 14th rev. (Noord-Hollandsche Uitgeversmaatschappij, 1978); J.E. Andriessen, 'Economie in theorie en practijk', 6th rev. (Elsevier, 1980).
7. J. Fedder (J. van Gelderen), Springvloed: beschouwingen over industrieele ontwikkeling en prijsbeweging, 'De Nieuwe Tijd', 18, 1913, pp.253-77, 369-84, 445-64.
8. Prosperitats und Depressionsperioden, in 'Der lebendige Marxismus', Festgabe zum 70 Geburtstag von Karl Kautsky, ed. O. Jensen (Jena, 1924); 'Het Economisch Getij' (J. Emmering, Amsterdam, 1929).
9. P.C. Harthoorn, 'Hoofdlijnen uit de ontwikkeling van het moderne bankwezen in Nederland voor de concentratie', thesis, Econ. Univ. Rotterdam, 1928.
10. 'Het terugvinden van het evenwicht', lecture, Barchem Circle, August 1935.
11. 'De konjunktuur' (Arbeiderspers, Amsterdam, 1933); Vertraggingsgolven en levensduurgolven, in 'Strijdenskracht door Wetensmacht', ed. J.v.d. Wijk (Arbeiderspers, Amsterdam, 1938).
12. Het raadsel van de lange golven, 'Economisch en Sociaal Historisch Jaarboek', 37, 1974, pp.269-84.
13. 'Die industrialisatie in Nederland tussen 1850 en 1914', history thesis, Free University, Amsterdam, 1968.
14. 'Landbouw en bevolking tijdens de agrarische depressie in Friesland (1878-1895 ', history thesis, Free University, Amsterdam, 1971.
15. See Wilhelm Abel, 'Agrarkrisen und Agrarkonjunktur in Mitteleuropa vom 13 bis zum 19 Jahrhundert' (Paul Parey, 1980); B.H. Slicher van Bath, 'The Agrarian History of Western Europe AD500-1850' (1963); B.H. Slicher van Bath, Agriculture in the vital revolution, 'Cambridge Economic History of Europe', 5, 1977, pp.43-132.
16. See for example James B. Shuman and David Rosenau, 'The Kondratieff Wave' (Delta, 1972).
17. For example W.W. Rostow, 'The World Economy: History and Prospect' (MacMillan, 1978).
18. See the revised edition of the 'Algemene Geschiedenis der Nederlanden', vols 12 and 13 (Unieboek, 1977 and 1978).
19. For instance Jean Gouvier, Feu Francois Simiand? in 'Conjuncture Economique - Structures Sociales', Hommage a Ernest Labrousse (Mouton, 1974), pp.59-78; also much of the research on innovations.
20. See Eric J. Hobsbawn, 'Labour's Turning Point 1880-1900', 2nd edn (Harvester, 1974).
21. 'Prosperitat und Depression', rev. edn (Bern, 1948), pp.329ff.
22. 'Das Problem der sozialokonomischen Synthese' (Lund, 1938), p.168.
23. 'Konjunkturzyklen I', translated (Vandenhoeck & Ruprecht, 1961), pp.154ff.
24. 'Handworterbuch der Staatswissenschaften', 4th edn, vol.VI (Jena, 1925).
25. Beitrage zur Erforschung der wirtschaftlichen Entwicklung', Heft 1 (Fischer, Stuttgart, 1957), pp.170ff.
26. Ibid., p.174.

27. Ibid., p.180.
28. Ibid., p.176.
29. Op.cit., p.xxiv.
30. 'Op het breukvlak van tween eeuwen', 2nd edn (Querido, 1976), p.49.
31. Op.cit., p.53.
32. See Thomas S. Kuhn, 'The Structure of Scientific Revolutions', 2nd edn (University of Chicago Press, 1975), p.85.
33. T.J. Broersma, 'De lange golf in het economisch leven', economics thesis, University of Groningen, 1978, p.30.
34. See note 14.
35. See note 5; Patrick O'Brien and Caglar Keyder, 'Economic Growth in Britain and France 1780-1914' (Allen and Unwin, 1978), p.60.
36. See Shepard B. Clough, 'European Economic History', 2nd edn (McGraw Hill, 1968), pp.389-90.
37. For example Hans Jurgen Puhle, 'Von der Agrarkrise zum Prafaschismus' (Franz Steiner, 1972).
38. Knut Borchardt, 'Wandlungen des Konjunkturphanomens in der letzten hundert Jahren' (Munich, 1976); Knut Borchardt, Wachstum und Wechsellagen 1914-1970, in 'Handbuch der deutschen Wirtschafts- und Sozialgeschichte', band 2, ed. Wolfgang Zorn (Ernst Klett, 1976), pp.685ff.
39. Van Gelderen already saw 1911 as a turning point 'Springvloed', p.17).
40. See F.A.G. Keesing, 'De conjuncturele ontwikkeling van Nederland en de evolutie van de economische overheidspolitiek 1918-1939', 2nd edn (Sun, 1978), p.23.
41. For Germany see Dietmar Petzina and Werner Abelshauser, Zum Problem der relativen Stagnation der deutschen Wirtschaft in den zwanziger Jahren, in 'Industrielles System und politische Entwicklung in der Weimarer Republik', ed. H. Mommsen, D. Petzina, B. Weisbrod (Droste, 1974), pp.57-75.
42. By a project group of the Research Group Long Term Fluctuations.
43. Otfried Hatzhold, Geburtenruckgang und Wirtschaftspolitik - Gedanken zu einem demookonomischen Paradoxon, 'Wechselwirkungen zwischen Wirtschafts- und Bevolkerungsentwicklung' (IfO-Studien zur Bevolkerungsokonomie 1, 1980), pp.262-274.
44. NRC-Handelsblad.
45. W.F. Wertheim, Het contrapunt in de samenleving, 'Weerklank op het wek van Jan Romein' (Wereldbibliotheek, 1953), pp.210-17.
46. See a report by D.J. Meuzelaar.
47. See Julius Shiskin, 'Signals of Recession and Recovery' (NBER, 1961); Philip Klein and Geoffrey H. Moore, 'The leading indicator approach to economic forecasting - retrospect and prospect' (NBER working paper no.941).

4 Technology and conditions of macroeconomic development

some notes on adjustment mechanisms and discontinuities in the transformation of capitalist economies

Giovanni Dosi

This paper concerns the role of technology and innovation in shaping the pace and direction of economic change. In particular, we will address the following question: are there some fundamental characteristics of technical change which can account for long-term cycles and/or long-term changes in the rate of macroeconomic growth? In order to develop our argument, we shall briefly discuss some fundamental features of technology and the mechanisms of interaction between the latter and the economic environment. We will then develop the argument in relation to the power and scope of the homeostatic forces present in the capitalist system and suggest that technology is likely to play an important (although by no means exclusive) part in the explanation of long-term changes in the rate of macroeconomic activity. This paper is very impressionistic and relies very little on systematic empirical analysis. Its prime aim is to discuss a theoretical framework whereby we can account for the complex set of interactions in which technology plays a part.

It is an evident stylized fact of modern economic systems that there are forces at work which keep them together and make them grow despite rapid and profound modifications of their industrial structures, social relations, techniques of productions, patterns of consumption. We must better understand these forces in order to explain possible structural causes of instability and/or cyclicity in the performance variables (such as the rates of macroeconomic growth). The relative dynamic stability of industrial economics and their regularities were what classical economists found fascinating and worth studying - I must confess that I, too, find that same object fascinating and relatively unexplored since then.

It might be useful to start from a more explicit definition of 'dynamic stability' and homeostasis. We probably live in the first social structure where technological, social and economic change is a fundamental feature of its functioning. For the first time, what we could call the 'bicycle postulate' applies: in order to stand up you must keep cycling.(1) However, changes and transformation are by nature 'disequilibrating' forces. Thus, there must be other factors which maintain relatively ordered configurations of the system and allow a broad consistency between the conditions of material reproduction (including income distribution, accumulation, available techniques, patterns of consumption, etc.) and the thread of social relations. In a loose thermodynamic analogy, it is what some recent French works call 'regulation'.(2) This important concept has clearly nothing to do with the meaning in which it sometimes is used by American neo-classical economists, but hints at the presence in the system of inner adjustment processes, institutions and servo-mechanisms which are vaguely similar to a thermodynamic system of engines and thermostats, whereby the structures of feedbacks prevent it from overheating, exploding or stopping. The problem of long-term cycles or, in any case, changes in the rates of macroeconomic activity pertain precisely to this level of analysis: are there structural features which produce crises in the regulatory set-up so that the engine tends to stop or to explode?

Let us heroically subdivide the overall socio-economic fabric into three domains, which we will call (i) the system of technologies, (ii) the 'economic machine', (iii) the system of social relations and institutions.

These three domains clearly interact with each other. Our analysis will start from the following hypotheses:

1. Despite powerful interactions, each of these three domains has rules of its own which shape and constrain every inducement and adjustment mechanism between them.
2. There is a limited number of configurations of these three domains which allows a relatively 'well-regulated' and smooth consistency between them. These define, so to speak, the 'possible worlds'.
3. Unbalanced or 'crisis' configurations do not necessarily also embody the necessity of the transition to other (more balanced or 'smoother') ones.

We shall clarify these points in relation to the interaction between the 'system of technologies' and the other two.

Some properties of technical change

I firmly believe that the idea of 'production possibility sets', with the corollary of technology as a malleable and reactive black box, to use Rosenberg's terminology,(3) has been one of the most damaging products of the neo-classical approach to economic theory. It has been developed without reference to any 'stylized fact' of the production process or the historical pattern of technical change, but it is a core assumption of the neo-classical theory of value which - in its clearest form - stands or falls with it.(4) Thus we should not be very surprised that, over the last hundred years, mainstream economists have been relatively more interested in the possible existence of contingency markets for red herrings than in the overall patterns of technical change. However, in the past two decades, a few important contributions have revived what we would call, 'grosso modo', a classical approach to technical change. We must mention, among others, Freeman, Nelson and Winter, Rosenberg, and Sahal. Along these lines, in another work,(5) we argue the general existence of 'technological paradigms' and 'technological trajectories'. What these concepts try to capture are a few fundamental properties of technology and innovation which are worth recalling.

a. The 'system of technologies' (techniques of production, nature of the products, the processes of their generation, the directions of progress, etc.) embodies rules and procedures which, at least in the shorter term, maintain some degrees of autonomy with respect to the forces of economic inducement.
b. Technical progress is a strongly selective activity which embodies normative rules similar to positive and negative heuristics (telling the directions where to go and where not to go).
c. Technical change can be distinguished between the emergence of new technological paradigms and 'normal' technical progress along the technical and economic dimensions of the 'trajectories' which a paradigm defines.
d. The process of generation and selection of new technological paradigms does not depend only upon economic factors, but also the timing and the nature of scientific advances and 'lato sensu' institutional factors.
e. Economic inducements, in the form of changing patterns and levels of demand, income distribution, and relative prices, are clearly powerful forces which shape the directions and the rates of 'normal' technical progress. Moreover, they may represent an incentive or an obstacle to the emergence of new paradigms, but they are not by any means a sufficient condition for that emergence. Finally, it must be stressed that even the inducement mechanisms of normal technical progress occur within the boundaries defined by the paradigms and the trajectories.
f. The distinction between major changes in technological paradigms and 'normal' progress broadly corresponds to the distinction between continuities and discontinuities in the process of technological advance.
g. The nature of technological progress is such that, regarding process innovation, new techniques are generally superior to the old ones irrespective of income distribution.(6) This does not mean that changes in distributive shares do not affect the direction of change.

They do affect it (within the mentioned boundaries). However, the 'better' techniques that are at last developed would have been adopted even at the 'older' income distribution. This introduces important irreversibility properties into the system.

h. As highlighted by Rosenberg,(7) there are often complementarities and interdependencies between different technological paradigms and trajectories. The phenomenon provides a thread of technological stimuli, incentives, bottlenecks to overcome, opportunities to sieze, etc. Even if all these mechanisms are ultimately related to the economic use of technologies, they contribute to provide an inner momentum to technical change which is not simply 'reactive' to the economic environment.

All these considerations taken together lead us to conclude that the adaptiveness of the technological system to a given economic and social environment is bound and limited. Conversely, each state of the 'technological system' defines a relatively limited set of possible macroeconomic conditions and social relations. In order to obtain an intuitive idea of these two points, the reader - through some reductio ad absurdum - may think of the enormous social need of labour-saving innovations in Florence at the end of the 15th century or the desperate requirement by the Roman Empire of a modern telecommunications system. Obviously, neither of them was accomplished. The serious point is that the homeostasis between the technological system and the other two defined above is only a limited one. Thus, this allows the possibility of major discontinuities, crises and fluctuations in the economic system itself.

We must now push the analysis a step further and explore whether there are factors at work which may also help in producing these discontinuities.

On the duality of technical change

Technical change is fundamentally about two things: either producing existing commodities or services with fewer inputs (i.e. more efficiently), or producing new commodities or services. In practice, product innovations of one sector are often process innovations for other sectors which are using them. The distinction, nonetheless, is theoretically fruitful. We just mentioned that process innovations necessarily imply some input saving. We can be more precise and suggest that in capitalist economies where conflict over labour processes, income distribution and power and structural features, labour saving must be one of the fundamental dimensions of most technological trajectories. Moreover, any labour saving 'upstream' (i.e. in the production of commodities which are also productive inputs) represents an input-saving, in value terms, 'downstream'.

Finally, let us just mention our belief (which we share with classical economists) that developed industrial systems are (i) functionally characterized - in normal circumstances - by reproducability and not scarcity, (ii) demand-pulled in terms of macroeconomic activity, and (iii) balance-of-payment constrained. Under these conditions, paramount importance must be given to the broad duality of technical change which on the one hand continuously saves labour[8] and, on the other hand, creates new markets or expands existing ones by means of changing costs and prices of each commodity and service. The balance between demand creation[9] and labour displacement defines the endogenously generated rates of macroeconomic activity and utilization of the labour force. Amongst recent economic contributions, the issue is analysed in depth by Pasinetti,[10] who defines the general conditions of macroeconomic stability in a multi-sector model characterized by technical change. In terms of interpretation of long waves in economic activity - as we shall see below - Freeman, Clark and Soete[11] focus on changing patterns of this duality of technical innovation.

Notably, prevailing economic theory can assume that a compensation effect between the two fundamental results of technical change does generally exist owing to the properties of the duality price/quantities in general equilibrium frameworks. We already had the chance above of criticizing one of the core assumptions of that theory, namely the idea of production possibility sets. Symmetric with that, and equally essential, there is another core assumption based on consumption tests, characterized by wide substitution possibilities and well behaved with respect to relative prices.[12] We believe that this second core hypothesis does not generally hold, either. Let us suppose, instead, that:

(i) there are relatively stable 'baskets of consumption', which depend on income levels and distribution, and institutional factors (the modes of social organization, the culture of social groups, etc.);
(ii) the prevailing force of change in these baskets of consumption are changes in per capita income and in cultural traits, so that long-term consumption patterns broadly follow Engel's curves;
(iii) changes in relative prices do not induce generalized substitution, but primarily affect the form and steepness of Engel's curves themselves[13] (in regard to final goods) or the process of diffusion of innovations (in regard to producer goods).

If these alternative hypotheses on consumption patterns are adequate 'stylized facts', then duality of technical change becomes a crucial knife edge in modern growth. The homeostasis between productivity effect (labour saving, increasing division of labour, economies of scale, learning-by-doing, etc.) and demand effect (creation of new commodities and markets, induced demand by means of decreasing prices of the commodities in terms of income, and more generally endogenously generated growth of income)(14) cannot be guaranteed ex hypothesi.

Freeman, Clark and Soete(15) argue that a fundamental mechanism which generates long waves is the changing balance between product and process innovations. In another work(16) we try to prove on theoretical grounds that Freeman-Clark-Soete's hypothesis has rather general macroeconomic implications: other things being equal, process innovations have a lower impact on aggregate demand than equivalent product innovations.(17)

An enormous amount of work still has to be done in order to develop multi-sector disequilibrium models embodying reasonable assumptions on the nature of technical change and the patterns of consumption. Moreover, it is still extremely difficult to link the foregoing considerations with aggregate growth models, roughly speaking, in the post-Keynesian tradition.(18) In any case, the appreciation of the possibility of imbalances between the dual effects of technical change allows the identification of what we could call the Keynesian gap, i.e. the difference between endogenously generated rates of macroeconomic activity and those which would correspond to the full employment of the labour force. (19) In order to prove that these Keynesian gaps are possible and likely in the long term, we need to follow a few theoretical steps, namely:

a. Are there major discontinuities in technical innovation, so that new paradigms operate a clustering effect upon bunches of innovation?
b. Can one define long-term shifts between process and product innovations, capable of affecting the balance between the dual features of technical change?
c. Even if the answer to the previous point is a positive one, are there compensating mechanisms within the 'economic machine', so that long-term changes in productivity need not show in changes (with the opposite sign) in the rate of macroeconomic growth and/or of the utilization of the labour force?

For the time being we will limit the discussion to a closed economy context. In a later section of this paper we will introduce the incentives and constraints provided by international interdependencies.

Regarding question a, the foregoing discussion on the technological paradigms, the likely interdependencies between different technological trajectories, the partly exogenous timing of paradigm generation, etc. can be considered as a strong argument supporting the hypotheses of discontinuities and clustering in technological innovation.(20) We shall focus instead on questions b and c.

A few hypotheses on productivity trends, consumption patterns and overall regulation

We have already mentioned that the assessment of the balance between product and process innovation is very difficult in practice, since outputs of some sectors are often inputs for others. From a theoretical point of view we can define two pure 'ideal types'. First, a pure process innovation is one which affects the input coefficients of any one vertically integrated sector(21) defined by the manufacturing of an existing commodity without creating any new vertically integrated sectors or directly affecting the levels of demand of other existing ones.(22)

Conversely, a pure product innovation is that which creates ex novo a new vertically integrated sector without substituting for existing ones. In actual fact, most innovations can be placed on a continuum between these two extremes. However, the net balance of their dual impact on aggregate levels of demand will depend, ceteris paribus,(23) on their nearness to either extreme.(24) There are, in our view, two reasons why there is likely to be a long-term changing balance between process and product innovations. The first is, loosely speaking, 'natural', in the sense that it is related to the inner features of technological trajectories. It is likely that the mechanisation of the processes associated with the production of innovative commodities and their use will take time and require the technological paradigm to be relatively established. Moreover, economies of scale in production will be exploited only beyond a minimum scale of production and a minimum level of standardization of the products. Finally, the range of new products which can be developed on the ground of a given technological paradigm is limited by both the nature of the paradigm itself - including the physical limitations and trade-offs it embodies - and by the 'stickiness' of consumption baskets, in the case of the final goods. (We shall come back to this point below.) These characteristics of technological paradigms and trajectories are clearly consistent with Abernathy and Utterback's(25) and Freeman-Clark-Soete's(26) analyses of technological patterns.

There is another set of macro-social and macroeconomic factors which can account for long-term changes in the process-bias (i.e. an increasing productivity effect) of technical change. To repeat, inducement mechanisms, from economic signals to the directions of technical change, are always at work, within the boundaries defined by the trajectories of normal technical progress. It is likely that periods of sustained rates of macroeconomic growth with associated low levels of unemployment exert a structural influence on income distribution, increasing the wage share in income, and on the levels of industrial conflicts, related to the modes of the labour process and, lato sensu, to power. These issues, in the long-wave context, are analysed in

Technology and conditions of macroeconomic development   67

depth by Salvati(27) who discusses Kalecki's and Phelps Brown's theories and by Screpanti(28) who develops an exciting model of long waves related to working-class conflicts and insurgencies. With regard to the issue we are discussing here, it is plausible to assume that increasing wage shares and increasing levels of conflict will stretch paradigms and trajectories to the limit in order to find labour-saving techniques.

Even these scattered remarks allow us to see that the long-term changes in the net balance between the dual properties of technical change is nothing but the 'economic tip' of a much greater set of functional relations which hint at the requirement of consistency between the 'technological system', the institutional set-up and the economic machine, which we mentioned at the beginning of this paper. More precisely: the dual economic features of technical progress - which determine the endogenously generated levels of aggregate demand - are affected by the pattern of consistency (or the 'mismatching') between:

a. the nature of the fundamental technological paradigms;
b. the nature of production and labour processes associated with them;
c. the rules of the game between the major social groups (which, notably, is a fundamental force in determining levels and changes in productivity - for given techniques - and income distribution);
d. the baskets of consumption, which, as already mentioned, are a function of income levels, income distribution and, given the latter, of the ways a society organizes the use of non-working time, the provision of services, etc.

We may illustrate this point with extremely 'unscientific' and impressionistic remarks on the post-war 'golden age' of development.(29) On the technology side, the sustained rates of growth were based on the rapid growth of a few fundamental technologies, e.g. automobiles, petro-chemicals, electrical consumer durables, capital equipment related to mechanized mass-production and 'Tayloristic' productive processes.(30)

On the labour-process and institutional sides, the 'Tayloristic' workers have been, despite profoundly different institutional arrangements across countries, one of the fundamental constituents of some kind of 'corporatist bargain' involving political commitment to full employment by the major western governments, a more or less explicit indexation of real wages on productivity increases, a relatively large control of corporations on labour processes guaranteed by the control of the trade unions upon the shop floor.(31) These are broadly the features of the overall 'monopololistic regulation' of the systems analysed by Aglietta, Boyer, and Mistral.(32) On the consumption side, the 'baskets' for the majority of the population in developed economies were enlarged to include new durables and/or the substitution of durables for traded services, as argued by Gershuny(33) (e.g. substitution of automobiles for public transportation, etc.). Notably this enlargement was allowed by both rising income levels and the 'corporatist arrangements' on real wage rates.(34)

Contextually, the smooth working of the forces endogenously generating aggregate demand within the 'economic machine' were allowed, among other things, by relatively low levels of industrial conflicts: what are sometimes called 'optimistic animal spirits' implied relatively high levels of investment generated both via the accelerator and autonomously, in relation to the opportunities offered by technical progress and new potential markets. 'Well-regulated' patterns of consistency between the three fundamental systems (technology, economics and institutional frameworks) can generate 'virtuous circles'. However, one can also see how the golden-age conditions can reach critical thresholds and, so to speak, 'cooling feedbacks'.

Again, let us just make a few scattered and impressionistic remarks on the recent crisis period. Even leaving aside exogenous shocks, it is reasonable to assume that inner trends have been at work, making the continuation of golden-age conditions increasingly difficult. For example:

(i) The increasing levels of industrial and political conflict around the end of the sixties must have had some effect on the rate of investment, even if we hold the innovation opportunities constant.(35)

(ii) There certainly is some truth in the suggestion that market saturation started to appear for the most dynamic items of final demand (e.g. consumer durables such as cars, brown and white goods, etc.). This is obviously not because people have all their 'needs' satisfied, but significant additions to the basket of consumption (e.g. the 'electronic home', etc.) require deep and long-term changes in the cultural patterns of use of time, the modes of social organisation, etc. At the same time, non-marketed(36) and/or 'positional' services have become increasingly important. The problem is that most of these services could hardly become an 'engine of growth' because they are not available under capitalist conditions of production of constant/increasing returns. Some of them are intrinsically 'positional'(37) and scarce (such as spending a vacation in the Galapagos). Others have a politically-determined scarcity (such as health, education, etc.). Neither can experience for the time being the virtuous circle conditions between increasing demand and increasing supply productivity which characterize manufacturing production.

(iii) Both technical progress on existing technologies and industries (such as cars) and the emergence of new technologies and new industries (such as microelectronics) have undermined the size and strength of the social constituencies which were behind the Keynesian-corporatist social bargain of the 'golden age'. This has happened both to the composition of industrial capitalism and to the structure of the working class. Think, for instance, of the changing balance of power in the USA between the old unionized companies on the East Coast and Great Lakes, and the new West Coast and Southern companies. More generally, one can find an archetypical illustration of this point in the ebb of the political and contractual bargaining strength of auto workers from the sixties to the eighties, throughout OECD countries.

These few notes cannot provide any analysis of the transition from post-war high-growth conditions to the present crisis. We just wish to illustrate how trends endogenous to the relationship between technological conditions, economic forces and institutional arrangement may lead to crisis feedback or, at least, a cooling down of growth patterns and increasing levels of social strain. The relative autonomy of the inner rules of the three mentioned fundamental systems is fully at work. The search for a new structural framework of regulation between the three is long, painstaking, often requires major crisis (and sometimes war), and success is not guaranteed. Certainly, the emergence of new fundamental technological paradigms changes the scope of 'possible worlds', and both demands and tends to induce far-reaching transformations in the economic structure and in social relations. However, I believe, technology alone can neither produce those qualitative transformations in the economy which guarantee long-term dynamism and stability of growth, nor adjust the modes of reproduction of societies structurally characterized by the possibility of conflict over labour processes, income distribution and, ultimately, power. To focus the imagination on this point, take the example of the 'Tayloristic' paradigm in mechanical manufacturing. Certainly it allowed the 'Fordist' pattern of industrial relations,(38) but by no means produced it, and even less did it produce Roosevelt and, later, the post-war macro-social and macroeconomic arrangements.

The foregoing discussion leads us to support the hypothesis that major discontinuities in both the overall functional relations between technology, economics and institutions and, more specifically, in the net macroeconomic impact of technical change, are possible and likely. However, must these discontinuities reflect themselves in long-term discontinuities in the rates of growth and/or of employment?

Adjustment mechanisms and rates of macroeconomic activity: is structural unemployment possible in the long term?

Even after allowing a changing macroeconomic impact of technical change, one must prove that these changes are sufficient to generate long-term changes in the rate of activity of the economic system. In other words, one must prove that, beyond short-term frictions, adjustment rigidities, etc. there is within the economic system an insufficient homeostasis so that there can be major discontinuities in the rates of investment and long-term lack of clearing of the labour market. This is a crucial theoretical point for any theory of long waves or long-term fluctuations, as we are reminded by Salvati(39) (1977), in the sharpest criticism of Freeman's theory of long waves that I am aware of, coming from within the same family of non-believers in Say's Law. On this issue, I will be able only to suggest some hypotheses and outline many unsolved questions in search for a sound theoretical treatment.

Following Salvati, let us recall the simple mathematics of Harrod's growth model:(40)

$$\dot{n} + \dot{\pi} = (I/Y)(Y/K)$$

where the dots stand for the percentage rates of growth and the symbols means the labour force (n), labour productivity (π), investments (I), income (Y) and the capital stock (K). The equation simply states a condition: if we want to have and keep full employment, the 'natural' rate of growth (that equal to the sum population and productivity growth), on the left-hand side, must equal the ratio of investment to income 32 times the inverse of the capital/output ratio (i.e. the 'warranted' rate of growth), on the right-hand side. Are there forces which can maintain the effective rate of growth at that level or push it there from initial conditions of disequilibrium?

The changing balance between productivity effect and demand effect, discussed in the previous section, implies both changes in π and (I/Y). For convenience, let us subdivide the total investment between that part (I_a) which is strictly autonomous, in the sense that it is related to levels of innovative opportunities which are exogenous to the macroeconomic forces we consider here, and the other part (I_i) which is macroeconomically induced via multiplier/accelerator mechanisms ($I = I_a + I_i$). Suppose productivity increases and innovation-related investment opportunities decrease. Then a gap between the warranted and the natural rates will emerge or will increase, other things being equal. Notably, our hypotheses on the nature of technologies and technical change rules out the possibility of the powerful compensation mechanism based on changes in the capital/output ratio, since the universe of best-practice techniques is likely to be very limited and there might well be inferior or superior techniques irrespective of income distribution. The irreversibility properties of technological dynamics show their effect. However, this is only the beginning of an answer. We may just mention a few important issues which, in my view, still remain open:

a. What happens to the effective rate of growth once we allow for reasonable assumptions on microeconomic behaviour and abandon the implicit one-sector one-commodity formulation of the model? Will the effective rate tend to converge on that warranted rate defined in the aggregate? Two ambitious and exciting models have recently been proposed, with different approaches, by Pasinetti(41) and Nelson and Winter.(42) Pasinetti essentially provides a system of reference for the changing equilibrium conditions which must be fulfilled in the presence of many sectors characterised by different rates of productivity increase and changing demand patterns. Behavioural assumptions, however, are absent and we do not know what are the actual paths of adjustment of the system. Nelson and Winter's evolutionary theory, on the contrary, is based on a complex set of microeconomic assumptions on firms operating in environments characterized by technical change. There, however, we face somewhat opposite difficulties in the sense that there is no easy link with macroeconomic variables such as income distribution or saving propensities at aggregate level.

b. What is the precise relationship between income distribution (and the related rate of profit) and the rates of investment?(43) The importance of that relationship should be clear. If the elasticity is high, then we are in the presence of a strong fluctuation-generating mechanism (both up and down). Take the example above, with increasing and decreasing innovation-related investment opportunities. Suppose, however, that a certain period of relatively high unemployment decreases the workers' bargaining power and with that also the wage share in income. Even if the capital/output ratio does not change, investment elasticity to the profit rate should increase the 'autonomous' rate of investment associated with given innovation-related opportunities. I will thus increase. What happens to the aggregate investment is more complicated since wages are also an item of aggregate demand, almost certainly with higher consumption propensities than profits: therefore a relative decrease in wages decreases the size of the multiplier and, through that, the accelerator effect on induced investments. A clear answer on the relationship between income distribution and rates of accumulation will not be given until we are able to assess, theoretically and empirically, the relative balance between 'autonomous' and 'induced' investment (as defined above) and the elasticity of investment to profit rates.(44)

These brief comments could only be headings of long analyses. What they show in any case is that, once we abandon orthodox general equilibrium hypotheses, no straightforward and evident homeostatic mechanism can easily be found to stabilize the growth rates in the long term around full-employment values.

Cycling together: dynamism and constraints of the international economy

Modern developed economies, it was hinted above, are 'demand-pulled' on macroeconomic grounds. When we allow for international trade, exports become a crucial item of aggregate demand. At the same time, imports, bearing a rather stable relation to income, constrain the maximum rate of growth compatible with a balanced foreign account. The relationship between foreign trade and domestic growth is sometimes expressed through the 'foreign trade multiplier'. (45) This is standard macroeconomics of the Keynesian tradition. We simply want to mention some implications of the patterns of technical change, suggested above, for the open-economy case. Technical progress is generally associated with (different degrees of) private appropriation of differential innovative capabilities as a corporate asset.(46) It is possible to show that such an appropriation also holds, by implication, for countries and not only for individual companies: other things being equal, the innovative process represents a divergence-generating mechanism, in terms of incomes and wage rates (in international currency), while, conversely, technological diffusion can be considered a convergence mechanism.

In another work,(47) we try to show how the rates of growth of each economy are constrained by the relative rates of innovation and/or imitation. There are rather strict conditions in the international arena which define the knife-edge of boundaries and stimuli to national growth possibilities. We can describe them impressionistically with the metaphor of two or more cyclists riding a tandem.(48) A priori, there is a wide range of velocities at which they can go. However, the speed at which each one of them cycles must be strictly coordinated. Otherwise, there are two possibilities: either the strongest one is capable and willing to pull the others, or the velocity is likely to adjust to the rhythm of the slowest one.

The countries on the technological frontier must be willing to bear the 'newcomer's burden', in two senses. First, they must provide a sustained rate of innovation which can induce in the long-term a sufficient stimulus for internal growth and, in the longer term, for international technological diffusion. The fulfilment of this condition alone, however, is by no means sufficient, and if the rate of innovation is higher than the rate of diffusion, it may even become perverse in the sense that it can induce divergence and increase the foreign balance constraints of late-coming countries. In these circumstances, the second requirement is that the country on the frontier (a) is successful in maintaining high rates of domestic macroeconomic activity (which obviously affect the foreign trade multipliers of the other exporting countries), and (b) provides a system of international payments and transfers on capital account which can finance the higher rates of growth of the late-coming countries. With these considerations in mind, one can appreciate the long-term implications of Mistral's hypothesis(49) that phases of economic growth are characterized by specific regimes of international economic relations and the dominance of one particular national economy.

There is an intuitive empirical reference in the British technological and economic hegemony until the 'great depression' in the second

half of the 19th century or the American technological and political leadership in the post-war 'golden age'. Conversely, periods of crisis and depression are generally characterized by the transition of the technological leadership between different countries(50) and the collapse of a regime of international payments (e.g. the crisis of the gold standard, or more recently the Bretton Woods system). Thinking of the role of technology factors in the recent crisis, one must mention:

a. the end of an era characterized by a generalized American technological advantage and fast rates of imitation by the rest of OECD countries and a few NICs;
b. the emergence, as a major technological and industrial power, of Japan, who, however, is unwilling and incapable of assuming the same role of sponsor of the international system which the US performed so well, both through the spontaneous behaviours of American multi-national corporations and an explicit policy of world political hegemony;
c. more generally, profound signs of technological divergence within the industrialized countries (both in terms of innovative capabilities and rates of productivity growth).

Also at this international level, a phase of 'virtuous circles' and smooth international consistency seems to have ended.

Some conclusions

I hope the sketchy hypotheses and remarks of this paper have helped to illustrate the role of technology in shaping and constraining the overall 'regulation' of the system. The presence of inner rules within the three fundamental domains of technology, economic mechanism and institutions (plus the set of international inter-dependencies provided by the international arena) accounts for (i) the existence of some degrees of independence of each of them; (ii) the lack of complete homeostasis between them; (iii) the possibility of both 'virtuous circles' and crisis configuration in their functional feedbacks. Our argument supports the existence of discontinuities of technical change itself (the emergence of new paradigms with their exogenous timing) and of discontinuities and mismatchings between technological paradigms, on the one hand, and forms of accumulation, patterns of consumption, institutional arrangements, on the other hand. The search for different configurations within the set of 'possible worlds' allowed by new technological paradigms is a difficult process, often ridden with breakdowns and crises.

In relation to the economic system, we tried to show the bound nature of the induction mechanisms from economic signals and incentives to technical change. Moreover, on the ground of given technological paradigms and trajectories, there may not be any economic adjustment mechanism powerful enough to compensate for long-term changes in the balance between the dual effects of technical change, which can at the same time enhance productivity and create new demand. Finally, technical change - in the form of different rates of innovation and diffusion of innovation between countries - contributes toward defining the changing patterns of competitiveness and thus the foreign trade multiplier and the balance of payment constraints that each national economy has to meet.

Throughout our argument, we have been referring to 'discontinuities' instead of the more precise and demanding term 'long waves'. Should these discontinuities, whose existence has been the thrust of this paper, also have a regular wave-like form? I must confess that I have much less interest in an exact periodization. The nature of the functional relations which can induce increasing strains on virtuous circles in the overall regulation of the system, and specifically in economic performance variables such as growth rates and unemployment rates,(51) is quite complex: believing that there are cycles of exactly the same length and of exactly the same form requires, in my view, a little bit of mysticism. The two arguments I know of that I found most convincing are the time-lags required in investment and scrapping decisions and, even more, the timing of the social learning and collective cultural change related to generational turn-over.(52)

Around fifty years is more or less the time span of two generations, which may embody the traits of different cultural attitudes, varying adaptability to different institutional arrangements, defeated hopes or rising expectations, etc.(53)

In any case, the search for a nicely fitted cyclical from should not obscure what I believe to be the core of the long-wave discussion: namely, the relationship between forces of inner adjustment and long-term discontinuities, within the economic system and between the economy, technological possibilities, and institutional set-ups.

Notes

1. I borrowed this expression, suggested in another context, from M. Salvati.
2. cf. M. Aglietta, 'A Theory of Capitalist Reputation', London, New Left Books, 1979; R. Boyer and J. Mistral, 'Accumulation, Inflation, Crises', Paris, Press Universitaire de France, 1978.
3. See N. Rosenberg, 'Inside the Black Box: Technology and Economics', Cambridge University Press, 1982.

Technology and conditions of macroeconomic development 75

4. I owe the full appreciation of this point also to the comments by M. Lippi. The role played by the idea of production possibility sets in the entire framework of neoclassical general equilibrium is masterly illustrated by F. Hahn (The Neo-Ricardians, 'Cambridge Journal of Economics', 1983), who develops a sophisticated and very subtle argument suggesting that (neo-)Ricardian theories of value and income distribution are a very special case of general equilibrium analysis.
5. G. Dosi, Technological paradigms and technological trajectories: a suggested interpretation of the determinants and directions of technical change, 'Research Policy', 1982. A different version with a discussion of the long-wave issue is in C. Freeman, 'Long Waves in World Economy', London, Butterworth, 1983.
6. We argue this point at greater length in Technological paradigms and technological trajectories, in C. Freeman, op.cit.
7. N. Rosenberg, op.cit.
8. Note that we are not talking here of possible labour-saving biases (i.e. whether technical progress is more labour-saving than capital-saving, etc.), but of the very fact that it displaces labour in so far as it increases labour productivity.
9. Clearly, we are talking of both demand for consumption goods and production inputs.
10. L.L. Pasinetti, 'Structural Change and Economic Growth', Cambridge University Press, 1981.
11. C. Freeman, J. Clark and L. Soete, 'Unemployment and Technological Innovation: A Study of Long Waves and Economic Development', London, Frances Pinter, 1982.
12. This can clearly be seen in general equilibrium analysis: the neoclassical duality between prices and quantities is founded on the symmetric possibility of substitution between techniques and between consumption bundles. The theory stands or falls with it, in the sense that both processes of substitution are necessary in order to yield the traditional properties of the model - in its simplest form - related to market clearing, relations between income distribution and marginal productivities, etc. Having, at least, substitution in consumption is necessary for the clearing properties of the model to hold.
13. A more thorough argument along similar lines in in L.L. Pasinetti, op.cit.
14. In this context we use the term 'endogenous' to mean both growth related to multiplier/accelerator effects and to innovation-induced investments.
15. Op.cit.
16. G. Dosi, 'On engines, thermostats, bicycles and tandems, or, moving some steps toward economic dynamics', Brighton, SPRU, University of Sussex, 1982.
17. By 'equivalent' we mean innovations that involve the same amount of capital investment.
18. A la Kalecki, Harrod, Kaldor, Pasinetti, J. Robinson.
19. Note that here we are completely neglecting the specific economic causes of the 'gap' and any normative issue related to whether simple 'exogenous' creation of aggregate demand is feasible and/or politically possible.
20. For in-depth discussions within the recent long-wave debate, about the clustering of innovations, see Freeman, Clark and Soete, op.cit., and J.J. van Duijn, 'The Long Wave in Economic Life', London, Allen & Unwin, 1983.
21. For a discussion of this concept, see L.L. Pasinetti, op.cit.
22. In the sense that process innovations are not associated with the increasing use of any input.
23. This assumption of the stability of all other economic conditions is clearly unrealistic. We shall discuss this precise issue below.
24. Again, for a formal demonstration, we must refer to Dosi, 'On engines.. ', 1982, op.cit.
25. W.J. Abernathy and J.M. Utterback, Dynamics of innovation in industry, 'Technology Review', 1978.
26. Op.cit.
27. M. Salvati, Political business cycles and long waves in industrial relations: notes on Kalecki and Phelps-Brown, in C. Freeman, op.cit.

28. E. Screpanti, Long economic cycles and recurring proletarian insurgencies, 'Review', 1983.
29. cf. M. Salvati, 'Sono possibili politiche economiche 'post-Keynesiane'? (ovvero@ sono i politici vittime di economisti defunti, o viventi?), Torino University, 1983.
30. cf. R. Coombs, 'Long Waves and Labour Process Change', paper prepared for the Conference on Long Waves, Paris, Maison des Sciences de l'Homme, 1983. He convincingly argues the importance on a macroeconomic level of capital equipment related to what he calls 'secondary' mechanization (as distinguished both from 'primary' mechanization, linked with steam engines and bench production, and 'tertiary' mechanization linked to automation and electronic machinery).
31. On the features of the 'golden age', see M. Salvati, 'Sono possibili.. ', 1983.
32. M. Aglietta, op.cit.; R. Boyer and J. Mistral, op.cit.; R. Boyer, Les transformations du rapport salarial dans la crise: une interpretation de ses aspects sociaux et économiques, 'Critiques de l'economie politique', 1981.
33. J. Gershuny, 'After Industrial Society, The Emerging Self-Service Economy', London, Macmillan, 1978; and Social innovation: change in the mode of provision of services, 'Futures', 1982.
34. An ambitious analysis of long-term changes in the mechanisms of formation adn macroeconomic role and wages is in R. Boyer, Les salaires en longue periode, 'Economie et Statistique', 1978, and R. Boyer, 1981, op.cit.
35. Cf. E. Screpanti, op.cit.
36. Cf. J. Gershuny, 1982, op.cit.
37. A la F. Hirsch, 'Social Limits to Growth', London, Routledge & Kegan Paul, 1977.
38. For a description and discussion, see R. Boyer, 1981, op.cit., and R. Boyer and J. Mistral, op.cit., and the bibliography quoted there.
39. M. Salvati, 'Technology, Long Waves and Structural Unemployment', paper prepared for the group on 'Science and Technology in the new socio-economic context', Paris, OECD, 1977.
40. Which is identical to the saving propensity, ex post by definition, and ex ante if we are in equilibrium (remembering that we are still considering a closed-economy case and we are neglecting government spending).
41. L.L. Pasinetti, op.cit.
42. R.R. Nelson and S. Winter, 'An Evolutionary Theory of Economic Change', Cambridge, Mass., Belknap Press, Harvard University, 1982.
43. On this issue, cf.: J. Mazier, B. Loiseau and G. Winter, Rentabilite et accumulation du capital dans les economies dominantes, 'Economie et Statistique', 1977; R. Boyer and J. Mistral, op.cit.; M. Aglietta, op.cit.; A. Lipietz, Conflicts de repartition et changements techniques dans la theorie Marxiste, 'Economie Appliquee', 1980; A. Lipietz, 'Derriere la crise: la tendance a la baisse du taux de profit', Paris, CEPREMAP, 1981; E. Mandel, 'Long Waves in the History of Capitalist Development', Cambridge University Press, 1980.
44. Of course, the question would become somewhat simpler if one fully accepted a 'Cambridge' income distribution, whereby it is the rate of growth which determines the rate of profit. In that case, however, we run into some difficulties in explaining the microeconomic conditions consistent with that hypothesis. Moreover, we feel quite uneasy in ruling out any institutional force as determinant of income distribution.
45. If we consider exports as the only 'autonomous' item of demand, then we can write:

$$Y = \frac{1}{m} X$$

where m=import propensity in income, Y=income, X=exports. Considering two autonomous items (exports and innovation-related investment), and starting from the accounting identity

$$Y = C + X + I_a + Ii - M$$

dividing by Y and re-arranging, we obtain

$$Y = \frac{1}{s + m - \lambda v} (X + 1_a)$$

assuming that the ratio of induced investment to income (I_i/Y) is linear in the capital/output ratio (λv), and s=saving propensity (i.e. 1-C/Y).

Equation 4 simply states the 'propulsive' effect that both innovation-related investment and exports have on income. It is not possible to discuss here the conditions under which the income, so obtained, is an equilbrium one.

46. We study this point at length in G. Dosi, 'Technical Change and Industrial Transformation: the theory and an application in the semiconductor industry', London, Macmillan, 1983.
47. Cf. G. Dosi, 1982, op.cit.
48. Ibid.
49. Cf. J. Mistral, 'La diffusion internationale inegale de l'accumulation intensive et ses crises', Paris, CEPREMAP, 1981; and J. Mistral, Competitivite et formation de capital en longue periode, 'Economie et Statistique', 1978.
50. For an historical analysis, see Freeman, Clark and Soete, op.cit.
51. We would choose first of all these two indicators of the possible discontinuities and long waves.
52. Cf. Phelps-Brown quoted in M. Salvati, 1983, op.cit.,and E. Screpanti, op.cit.
53. See E. Screpanti, op.cit.

5 New evidence on the shift toward process innovation during the long-wave upswing

Rod Coombs
Alfred Kleinknecht

Recent research into long waves has produced a great deal of useful theoretical and empirical material which may contribute to a process of synthesis. The general tendency of the research has been to move away from rigid, determinist explanations of long waves, and to incorporate a wide variety of interacting variables into multi-causal models. These variables fall into three very broad headings. First, there are the 'economic' variables such as rates of change of productivity, profit, growth, employment, etc. Analysis of these parameters at both aggregate and disaggregate levels is proving very useful (Freeman et al., 1982; van Duijn, 1983). Second, there are the 'institutional' variables, such as legal and political stimuli and constraints relating to labour markets and company practice; regional variations in the basic economic processes of the long wave; changes in international currency and trade institutions, etc. (Gordon, Edwards and Reich, 1982). Finally, there are the 'technological' variables, such as number, type and frequency of innovations, inventions and patents; and diffusion data on particular innovations or groups of innovations (Freeman, et al., 1982; Mensch, 1975; Kleinknecht, 1981).

The assembly of these three groups of variables into a coherent explanation of the long-wave phenomenon is still a long way off. So far, the lower turning point of the wave is still proving the most difficult and the most crucial part of the problem, and we do not address this topic in this paper. In general, more progress has been made in explanations of what happens once a long boom has started, and as it progresses to a long depression. However, data to test these explanations are still sparse.

There is then a pressing need for more empirical work on the technical, economic and institutional aspects of long waves in order to test the variety of competing theoretical accounts. Such empirical work must of necessity proceed slowly from one part of the problem to another and so on. In this paper, we make no apology for concentrating only on the technological variables, and in particular on innovation data. Due to the time period of our data, we concentrate on the post-war upswing. The analysis does have some implications for the lower turning point, however, and these are discussed in the conclusion.

The problems with the analysis of technological variables are both empirical and theoretical. Statistics are hard to obtain and invariably poorly suited to the task at hand. Despite considerable discussion, there is still no consensus on whether to concentrate on innovations or on diffusion processes. These difficulties have been usefully summarized by Freeman, Soete and Townsend (1982). Nevertheless, there is one proposition concerning the role of technical change during the long wave which is now the subject of some agreement, namely that there is a shift in emphasis from product change to process change as the wave progresses. This shift is related to the process of maturation of new industries which grow with the long-wave upswing. It is also related to the dynamics of the labour market during the long wave. This possibility is examined by Freeman et al. (1982) who suggest that labour shortages in the upswing may contribute to the pressures to mechanize. It is also consistent with the diffusion data and arguments concerning automation (see Mandel, 1975; Coombs, 1983).

Empirical evidence on a shift from product to process innovation has been reported for the case of a single industry several times. Freeman, et al. (1963) reported this for the chemical industry, and Abernathy and Utterback (1975) have incorporated the idea into their model of industrial development. More recently, Mensch (1976) has spoken of a shift from 'expansionary' to 'rationalizing' innovations as the long wave progresses. Freeman, Soete and Townsend (1982) believe that their own data, and those of Kleinknecht (1981) show the shift even more clearly.

In this paper, we use a new set of data (see below for description of sources) to test this proposition more rigorously for the period 1953 to 1973. There are three major differences between the analysis in this paper and that in the previous attempts to test this hypothesis:

1. As Freeman, Soete and Townsend have noted, the classification of innovations as product or process presents many difficulties. A new method of classification is described below.
2. Previous tests have been at a high level of aggregation, which may have obscured some information. In this paper we conduct the analysis first for the entire sample, which covers the whole of industry, and then for a sub-group of the most innovative and most rapidly growing industries. The results are quite different.
3. There must be some doubt about the statistical significance of the shifts reported in previous work. For example, in the paper by Freeman, et al. (1982) the shift in the share of process innovation of around 7% between the fifties and the sixties is based on a total sample for the two decades of only 85. The sample used in this paper is 500 for a similar period. It is therefore possible to assign

the innovations to individual years and analyse the significance of any shift by simple regression.

The data

The data are taken from a report prepared for the United States National Science Foundation by Gellman Research Associates (1976). The aim of this work was to collect a sample of 500 product and process innovations that embody significant technological change. The sample was restricted to innovations that were successfully introduced into the market during the period 1953 to 1973 (inclusive) in six Western countries: USA (63% of the sample cases), United Kingdom (17%), Federal Republic of Germany (7%), Japan (7%), France (4%), and Canada (2%). It was the intention of the authors to cover innovations from a broad spectrum of the economy. The sampling process started with compilation of a preliminary list of 1160 innovations obtained by a survey of trade literature published from 1953 to 1973.

The selection of innovations to be included within the final sample of 500 cases was performed by an international panel of seven experts. These panelists were encouraged to suggest any innovations for possible later inclusion which were not on the original list of 1160 innovations. The panelists suggested some 150 additional innovations. The resulting list of 1310 innovations was submitted to each of the panelists for ranking in order of importance. Those 500 innovations which received the highest ranking were then subjected to further investigation to ascertain dates, origins, etc.

The reconstruction of historical innovation data is a task with many difficulties and ambiguities. For instance, it is very hard to say how far the selection of innovation data from a literature survey imparts any bias, and it is not possible to judge the reliability of the decisions made by the panel of experts.

A sample of 500 innovations over a twenty-one year period is nevertheless a significant improvement on what has previously been available. Its randomness is certainly not worse than other samples and its size is an improvement on existing data.

Therefore, we assume that the 'Gellman sample' is a useful data base for examining major patterns of technological innovation on an international level during the period 1953 to 1973.

Classification principles

The objective of a classification scheme is to separate innovations that create new products or services from those that are aimed at producing the existing set of products and services in a more efficient way. However, as has recently been pointed out by Freeman, Townsend and Soete (1982), such a seemingly simple task is in fact very difficult. Besides the problem of how to deal with those cases that are somewhere between pure product innovations on the one hand and purely rationalizing innovations on the other hand, the standpoint of the observer is of some importance. From the perspective of an investment-goods producer, a new NC machine is a product innovation that may increase employment in his firm, whereas for the final user, the NC machine is a labour-saving process innovation. An automobile is an 'investment good' if used for business purposes, but it becomes a 'consumer good' if used for private leisure. It should be clear that for the purpose of long-wave analysis it is not very helpful to take the perspective of individual enterprises. Instead, we have to assess the place and function of a certain innovation within the macroeconomic production system; i.e. we have to take a macroeconomic viewpoint.

One approach would be simply to separate innovations introduced by investment goods producers and innovations introduced by consumer goods industries. As a third and fourth category we would have to leave out innovations coming from basic materials extraction industries that produce inputs for both consumer and investment goods industries, and innovations from sectors outside of the manufacturing industry (trade, transportation, government institutions, etc.) whose role is still obscure in a long-wave context. Although such a classification procedure relieves us from judgements about the character of individual innovation cases, it leaves us with several problems. First of all, it is by no means sure that investment goods industries only innovate new investment goods (i.e. process innovations) or that consumer goods producers only concenrate on innovating new final consumer products. Secondly, separation between investment goods and consumer goods producers is not always that clear; often the same enterprise is engaged in both types of production, and statisticians have to make pragmatic judgements about how to group it.

Nonetheless, this 'sector of origin' approach might yield some indication of how the relative innovative dynamics of consumer and investment goods industries develop over time. According to the hypothesis as outlined in the introduction, we would expect the investment goods producers to have a rising share over time of innovations in the sample, and the consumer goods industries to have an opposite trend.

In order to avoid the potential errors of the 'sector of origin' approach, we have developed a scheme within which to classify each innovation individually. This contains a number of categories which are shown in Figure 1 and described below, with some examples for each case.

First, there are the non-controversial cases of pure product innovations (P) such as colour TV, and pure process innovations (I) such as continuous casting of steel. Difficulties begin to appear with automobile-related innovations (seat belts, power steering, disc brakes or electronic ignition), since automobiles are also used for commercial purposes. Nonetheless, we decided to classify them as product innovations, arguing that cars are mainly used as private consumer goods. But what about 'computerized real estate marketing' or 'computerized passenger reservations for airplanes'? There is certainly an element of cost-reduction, but at the same time these systems are likely to offer new or improved services to consumers. The same is true of such cases as the 'world's first commercial jet aircraft', 'weather satellites' or an 'electron bombardment process to sterilize food and drugs'. None of these examples is a final consumer good. Nonetheless their main impact is providing new or improved products or services to final consumers. Since this type of innovation was quite frequently to be found in the sample, we decided to classify them as a separate class: investment goods aimed at providing new or improved products or services (IP).

Another special category was reserved for medical instruments and procedures (MED). In so far as medical apparatus is concerned, this category comes near to the above-named IP category. However, the MED category also covers services and know-how more directly related to the human body (improved heart pacemakers or procedures for transplanting human organs). We also included new pharmaceuticals in MED. On the whole, the MED category comes quite close to product innovations, although most of these innovations are not directly sold to final consumers.

Scientific instruments (SI) form another category. These innovations are used primarily in research laboratories, but to a certain extent also for industrial quality control. Therefore we conceive the SI category as coming relatively close to process innovations.

The sample contains two other types of innovation that are difficult to group into any of the categories so far mentioned. These are new technological devices (TD) and new technological materials (TM).

An example of TD is the laser, and an example of TM is epoxy resin. While these clearly are innovations, their function is not limited to one specific area, and generally this potential for multiple application is clear at the time of innovation. They therefore constitute new inventive inputs to other sectors outside their sector of immediate origin. These 'multi-purpose technologies' have a dual significance; they are innovations for the sector or firm which produces them, but they change the technological possibilities for the whole range of future product and process innovations. It would therefore be inappropriate to classify them as one or the other, since this would be to obliterate an important dimension of their role in the economy.

Having defined these categories of P, MED, IP, I, SI, TM and TD, it is possible to combine them in different ways to represent more or less inclusive definitions of product and process innovation. This is shown in Figure 1.

It will be clear from Figure 1 that we have decided to divide IP and T between the product and process innovations in the wide and extended definitions. This procedure avoids classifying these innovations wrongly to one side or the other, yet it avoids sacrificing the information contained in the distribution of these cases over time. The IP cases are not controversial in this respect. The inclusion of the T cases can also be argued on the basis that they may contribute to both product and process innovation. In any event, this scheme gives us a variety of ways of examining the proposed shift from product to process innovation, as well as the prospect of examining the T innovations separately.

Results

Numbers of innovations by sector of origin

This approach is based on a division of the economy into a service sector and a manufacturing sector. The manufacturing sector is further subdivided into three main branches:

* basic materials industries
* investment goods industries
* consumer goods industries

This subdivision follows that used by the DIW (German Institute for Economic Research). Grouping of the SIC sectors into these main branches is documented in Table A1, which shows the absolute numbers of innovations for the above-named sectors. Since the number of innovations in the total sample exhibits considerable fluctuation over time, it is better to express the innovations of each sector as a percentage of the total. The estimation of simple regressions over time reveals that the percentage shares in the sample of the service sectors and of the basic materials industries show no significant trend. The regression equations are as follows:

service sectors:
$y=23.516-0.300t$ (t-value:-0.697)

basic materials industries:
$y=27.715-0.181t$ (t-value:-0.520)

The investment and consumer goods sectors behave according to our hypothesis. Throughout the investigation period, the percentage share of innovations from investment goods producers shows a significantly

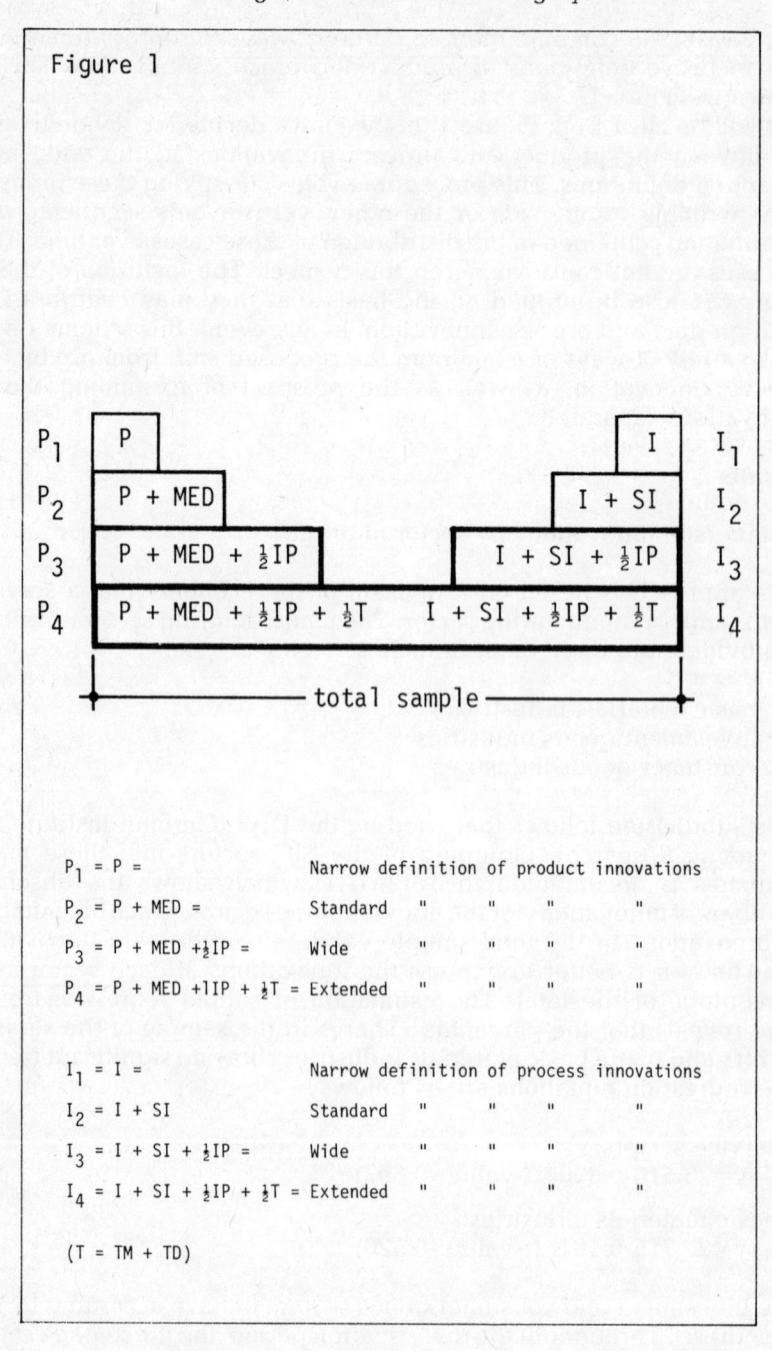

Figure 1

```
P₁    │ P                    │                          │ I              │ I₁
P₂    │ P + MED              │                          │ I + SI         │ I₂
P₃    │ P + MED + ½IP        │               │ I + SI + ½IP             │ I₃
P₄    │ P + MED + ½IP + ½T   │   I + SI + ½IP + ½T      │ I₄
```

|←————————— total sample —————————→|

$P_1 = P =$ Narrow definition of product innovations
$P_2 = P + MED =$ Standard   "   "   "
$P_3 = P + MED + \tfrac{1}{2}IP =$ Wide   "   "   "
$P_4 = P + MED + 1IP + \tfrac{1}{2}T =$ Extended   "   "   "

$I_1 = I =$ Narrow definition of process innovations
$I_2 = I + SI$ Standard   "   "   "
$I_3 = I + SI + \tfrac{1}{2}IP =$ Wide   "   "   "
$I_4 = I + SI + \tfrac{1}{2}IP + \tfrac{1}{2}T =$ Extended   "   "   "

(T = TM + TD)

increasing trend, and the share of consumer goods industries shows a reverse trend (see Graph 1). This can be taken as preliminary confirmation of the hypothesis that in the course of the post-war long-wave upswing, emphasis was shifting from product to process related innovations, i.e. the relative importance of the capital goods sector as a source of innovation was increasing, whereas the percentage share of innovations from the consumer goods industry was declining.

The results from the sample as a whole

The results of classifying the entire sample according to the categories discussed are shown in Table A2. Here again, we estimated linear regressions over time of the percentage shares of product and process innovations according to the various alternative definitions shown in Figure 1. The results are documented in Table 1.

Whether the definition used is wide or narrow, the slopes of the trends are quite small; i.e. the probability that the increases or decreases might be accidental is relatively high. Therefore the trends have to be interpreted with the utmost care. With the exception of I(1), the direction of the trends is consistent with the shifting hypothesis, but the significance of the slopes is too weak to take the results as a strong confirmation of the hypothesis. However, experience tells us that in innovation theory the most important information is often lost if we restrict our view to large aggregates. In the next section, therefore, we have further disaggregated the sample.

Disaggregation into modern and traditional sectors

In Kleinknecht (1981) the innovations of the Gellman sample were distributed into thirty sectors of the German manufacturing industry. This procedure was guided by the hypothesis that, given the strong integration of German industry into the world market, there should be a fairly good correlation between international innovation trends and industry growth patterns - if the Schumpeter approach is relevant. This correlation does indeed exist. The study reveals quite remarkable sectoral differences in the rates of growth of industrial production between sectors and shows that this corresponds with a one-tailed sectoral distribution of innovations. Taking into account differences in the rates of production growth as well as innovative behaviour, the study of Kleinknecht (1981) suggests that it is appropriate to separate the manufacturing industry roughly into two parts:

1. Highly innovative growth industries, which performed a locomotive function in the post-war upswing: chemicals, petroleum refining, rubber and asbestos, cars, aircraft construction, electrical equip-

Graph 1a: Annual Percentile Shares of Innovations from Investment Goods Industries in Total Sample

$$Y_t = 38.306 + 0.822 * year$$
$$t\ value:\ 2.620$$

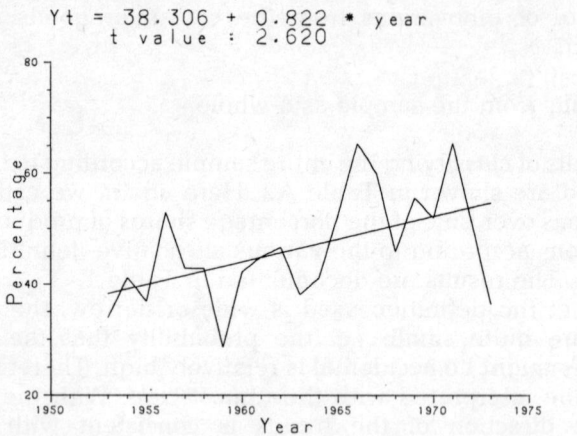

Graph 1b: Annual Percentile Shares of Innovations from Consumer Goods Industries in Total Sample

$$Y_t = 10.463 + -0.341 * year$$
$$t\ value:\ -1.847$$

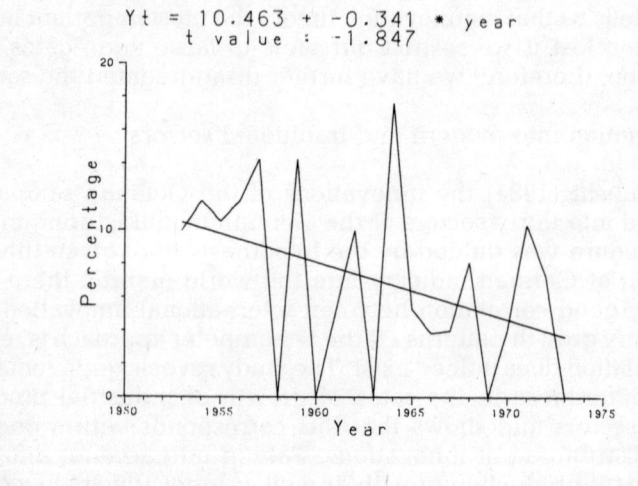

ment, precision engineering, plastics manufacturing.
2. Traditional industries, with more moderate growth rates and weaker innovation performance: mining, building materials, iron and steel, non-ferrous metals, sawmill and timber processing, woodworking, cellulose and paperboard, steel construction, machinery construction, shipbuilding, hardware and metal goods, fine ceramics, glass, wood manufacture, musical instruments/toys/jewellery, paper and board manufacture, printing and duplicating, leather manufacturing and processing, shoes, textiles, clothing, tobacco, food and beverages.

Graph 2 shows the numbers of innovations originating in these two parts of industry as a percentage share of the total number of innovations in the manufacturing industry (413 cases in all; i.e. 87 out of 500 cases fall outside the manufacturing industry).

The graph illustrates that during the post-war Kondratiev upswing there is a rising share in industrial innovation taken by the group of eight high-growth industries and, correspondingly, there is a considerable decline in the relative contribution of the older, traditional sectors. Let us now look at what is happening within these two groups, using the classification scheme presented above in Figure 1. A summary of the total period is given in Table 2.

The table demonstrates clearly the difference in the ratio between product and process innovations between the two groups. Independently of how we define product innovations (P, P + MED, P+MED+1/2IP+1/2T) it can be seen that the twenty-two traditional industries have very few of them. This implies that if there is any shifting from product to process innovations it can only have taken place within the modern industries; the traditional industries seem already to have shifted long before our observation period. This can also be seen in more detail from Table A3, which covers the same data on an annual basis.

Since there are so few product innovations in the traditional sectors (on any definition) it is not appropriate to pursue further the possibility of a product/process shift in these sectors. In the eight high-growth industries, however, it is possible to repeat the regressions on P1-P4 and I1-I4 to explore the shift within these industries.

The results are summarized in Table A4 and in Table 3 and Graphs 3 and 4. Table 3 shows that the regressions on the process innovations have significantly positive slopes on all four definitions (I1-I4). The slopes of the regressions on P1-P3 are not significant; only the slope on P4 is significant. It should be pointed out that our classification scheme and the nature of the data result in there being many more process innovations than product innovations in the sample (see Table A4). This inevitably makes the regressions on the product innovations less reliable. By the same token, however, we can have much more

Table 1

Regressions over time of percentage shares in total sample of product and process innovations (abbreviations and definitions: see Figure 1)

Product and process innovations according to different definitions	Regression equations	t-values
I_1 (narrow definition of process innovations)	$y_t = 51.331 - 0.017\,t$	-0.036
I_2 (standard definition of process innovations)	$y_t = 49.943 + 0.719\,t$	1.403
I_3 (wide definition of process innovations)	$y_t = 54.840 + 0.514\,t$	1.108
I_4 (extended definition of process innovations	$y_t = 68.160 + 0.452\,t$	1.198
P_1 (narrow definition of product innovations)	$y_t = 8.314 - 0.122\,t$	-0.523
P_2 (standard definition of product innovations)	$y_t = 13.624 - 0.185\,t$	-0.597
P_3 (wide definition of product innovations)	$y_t = 18.521 - 0.391\,t$	-1.066
P_4 (extended definition of product innovations)	$y_t = 31.840 - 0.452\,t$	-1.198

Table 2

Types of innovations by modern and traditional industries

	I	P	IP	T	MED	SI	D	Total
8 modern growth industries*	73 29.92%	34 13.93%	18 7.38%	70 28.69%	21 8.61%	21 8.61%	7 2.87%	244 100%
22 traditional industries**	115 68.05%	3 1.78%	13 7.69%	34 20.12%	1 0.59%	2 1.18%	1 0.59%	169 100%
30 sectors of manufacturing together	188	37	31	104	22	23	8	413
Sectors outside manufacturing industry (trade etc.)	59	1	2	12	5	6	2	87
Total sample	247	38	33	116	27	29	10	500

*Chemicals, petroleum refining, rubber and asbestos, vehicle construction, aircraft construction, car construction, electrical equipment, precision engineering and optics, plastics manufacturing.

**Mining, building materials, iron and steel, non-ferrous metals, sawmill and timber processing, wood-working/cellulose and paperboard, steel construction, achinery construction, shipbuilding, hardware and metal goods, fine ceramic, glass, wood manufacture, musical instruments/toys/jewellery, paper and board manufacture, printing and duplicating, leather manufacture, leather processing, shoes, textiles, clothing, food/tobacco/beverages.

Graph 2a: Annual Percentile Shares of Innovations from 8 Innovative Growth Industries in Total Sample

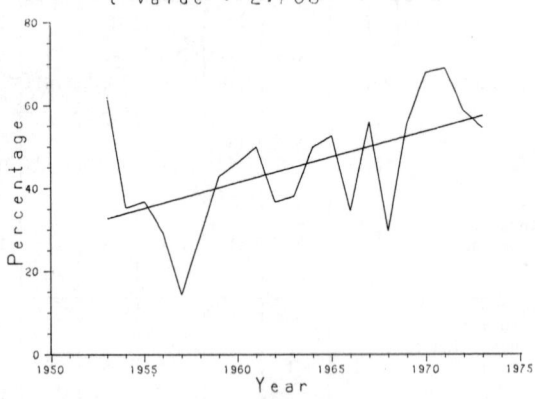

$$Yt = 32.830 + 1.235 * year$$
$$t\ value : 2.768$$

Graph 2b: Annual Percentile Shares of Innovations from 22 Traditional Industries in Total Sample

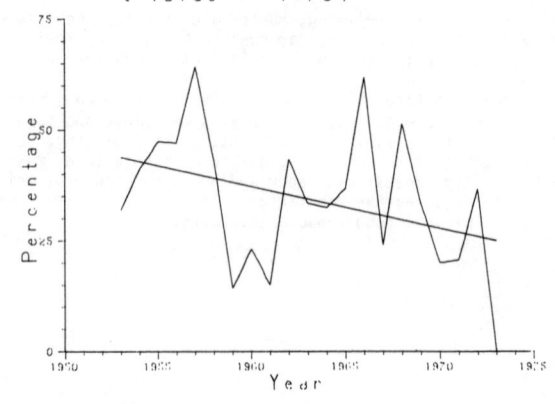

$$Yt = 43.688 + -0.935 * year$$
$$t\ value : -1.704$$

Table 3

Regressions on annual percentile shares of product and process innovations in total number of innovations within 8 modern, highly innovative growth industries (abbreviations and definitions: see Figure 1)

Product and process innovations according to different definitions	Regression equations	t-values
I_1 (narrow definition of process innovations)	$y_t = 23.903 + 0.911\ t$	1.724
I_2 (standard definition of process innovations)	$y_t = 24.483 + 1.682\ t$	2.911
I_3 (wide definition of process innovations)	$y_t = 30.027 + 1.384\ t$	2.554
I_4 (extended definition of process innovations)	$y_t = 50.904 + 1.007\ t$	2.007
P_1 (narrow definition of product innovations)	$y_t = 16.610 - 0.328\ t$	-0.649
P_2 (standard definition of product innovations)	$y_t = 22.674 - 0.333\ t$	-0.585
P_3 (wide definition of product innovations)	$y_t = 28.218 - 0.631\ t$	0.958
P_4 (extended definition of product innovations)	$y_t = 49.096 - 1.007\ t$	-2.007

Graph 3a: Annual Percentile Shares of Process Innovations (According to "Wide Definition") Within 8 Innovative Growth Industries

$$Yt = 30.027 + 1.384 * year$$
$$t\ value : 2.554$$

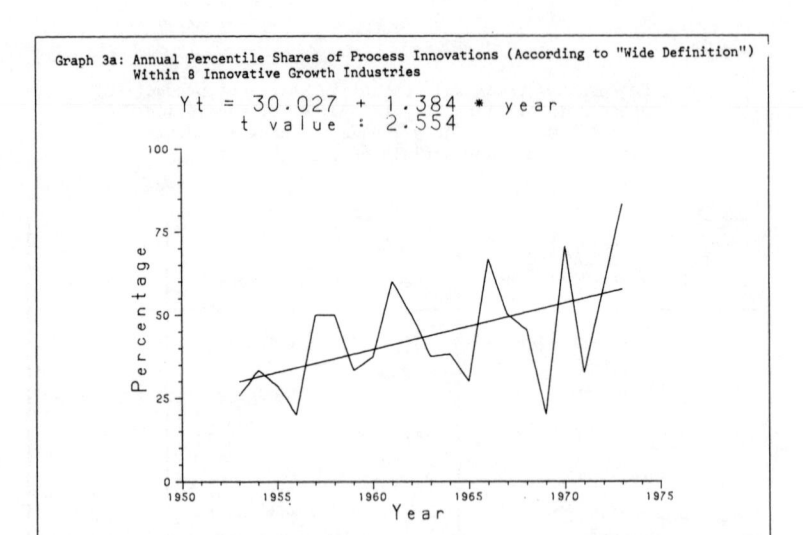

Graph 3b: Annual Percentile Shares of Product Innovations (According to "Wide Definition") Within 8 Innovative Growth Industries

$$Yt = 28.218 + -0.631 * year$$
$$t\ value : -0.958$$

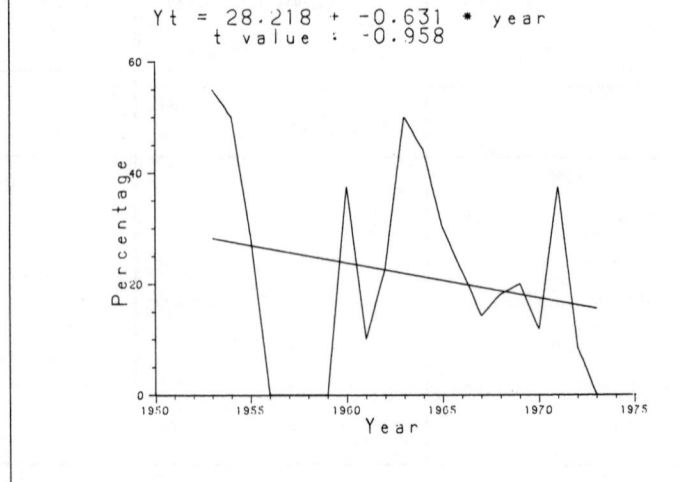

Graph 4a: Annual Percentile Shares of Process Innovations (According to "Extended Definition") Within 8 Innovative Growth Industries

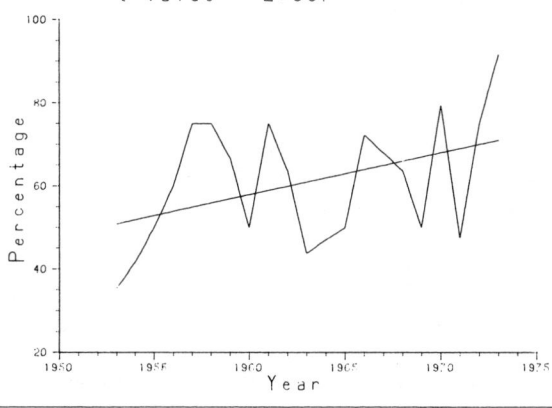

$$Y_t = 50.904 + 1.007 * year$$
$$t\ value : 2.007$$

Graph 4b: Annual Percentile Shares of Product Innovations (According to "Extended Definition") Within 8 Innovative Growth Industries

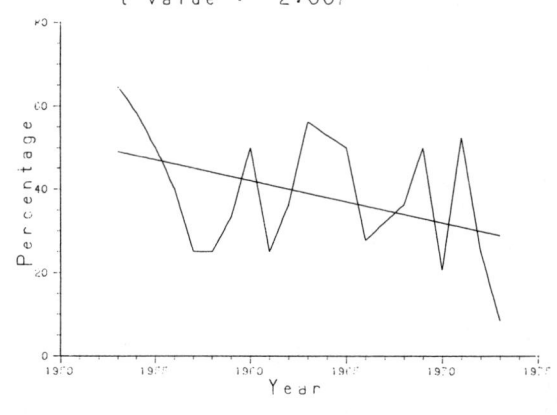

$$Y_t = 49.096 + -1.007 * year$$
$$t\ value : -2.007$$

confidence in the regressions on the shares of process innovations in the eight industries, and the latter are significant even on the most restrictive definitions (I1, i.e. primarily factor-saving investment goods). It does seem then, that 'rationalizing' innovations are becoming more important than 'expansionary' innovations in this sample as the upswing develops. Finally, it is important and interesting to note that the T innovations taken alone do not show any significant trend. (The regression equation for the percentile share of T innovations in total over time is: yt=40.645-0.664t (t-value: 1.000).

Conclusions

The first conclusion which can be drawn from these results is that the level of aggregation is very important in the study of changes in the character of innovation. The hypothesized shift from product to process innovation was not clearly visible in the aggregate sample except by means of the indirect 'sector of origin' approach. The individual classification of innovations did, however, show a strong confirmation of the hypothesized shift in the eight high-growth industries. This is an important confirmation and modification of the hypothesis.

Secondly, it can be concluded that this result gives further support to the models of long waves developed by the authors mentioned earlier, since it confirms some of the features of strong industry life cycles which play an important role in these models once the wave is under way. Furthermore, it is consistent with the arguments of Freeman et al.(1982) concerning the ability of the growth industries to continue to generate employment. It seems that as the wave progresses these sectors are likely to move toward 'jobless growth'.

If, in the economy as a whole, and in the growth industries of the upswing in particular, the rate of product innovations has fallen while the total rate of innovation has increased, then it is important to ask what the effect of various policies might be on that situation. Depressed demand will probably accentuate the trend, while expanded demand could not guarantee its reversal. Falling wages might reduce some pressure towards process innovation, but that does not in itself transfer effort to product innovation, especially if the falling wages are within a context of depressed demand. This suggests a role for technology policy, but raises the question of which technologies and products to promote; a familiar problem which will not be pursued further here.

On the difficult question of the nature of technical change at the lower turning point, the results are suggestive rather than strongly confirming any hypothesis. The fact that there was a higher level of product innovation at the beginning of the time period than at the end might possibly be consistent with the level also having been high during the latter phase of the long-wave depression. On the other hand, the peak may have come after the upswing began. The time period of the data (1953-1973) is just outside the period needed to examine this possibility. It is also important, however, that the T innovations are effectively random. This is suggestive of the possibility that fundamental technical changes may be less closely coupled to the long wave, as Freeman has suggested. Again, without more data we cannot be sure that there is not a peak or a trough in T in the downswing of the wave.

In any event, these results underline the importance of careful analysis of the nature of technical change in the long wave. It may be too crude to operate only with the categories of inventions and innovations, given the dual significance of the T innovations as multiple-purpose technologies. The tracing of connections between T innovations and their later role as inputs to other innovations would be a useful research strategy, especially if coupled to diffusion data on these and other innovations. In sum, there is a need to move beyond the simple counting of inventions and innovations, despite the suggestive results provided by the method to date.

Table A1

Annual numbers of innovations by main sectors of origin

	Non-manufacturing sectors	Manufacturing: Basic materials industries	Manufacturing: Investment goods industries	Manufacturing: Consumer goods industries	Total sample
1953	3	25	17	5	50
1954	4	4	7	2	17
1955	3	7	7	2	19
1956	4	2	9	2	17
1957	3	3	6	2	14
1958	4	4	6	0	14
1959	3	1	2	1	7
1960	8	7	11	0	26
1961	7	3	9	1	20
1962	6	7	14	3	30
1963	6	7	8	0	21
1964	6	6	14	6	34
1965	2	6	10	1	19
1966	1	7	17	1	26
1967	5	4	15	1	25
1968	7	10	17	3	37
1969	2	6	10	0	18
1970	3	8	13	1	25
1971	3	4	19	3	29
1972	2	15	21	3	41
1973	5	2	4	0	11
Totals	87	140	236	37	500
Corresponding SIC numbers:	154,161,162,173 374,401,422,431 442,452,461,478 481,483,489,491 494,495,508,602 632,737,739,769 806,891,892,919 951,957,962,966	109,121,131 324.325,327 329,331,339 333-335,281 282-287,289 291,301,242 261	344,351,371 353-359,373 372,376,360 362,365-367 369,381-384 386,341,347 349,342	321,243 249,393 262,307 311,222 228,231 203,206 208,209	

Table A2 Annual numbers of product and process innovations in the total sample (abbreviations and definitions: see Figure 1)

	Pure I	Pure P	I+SI	P+MED	IP	T=TD+TM	I+SI+½IP+½T	P+MED+½IP+½T	Totals
1953	16	7	16	15	10	9	25.5	24.5	50
1954	8	2	8	2	2	4	11	5	16
1955	13	2	13	2	0	4	15	4	19
1956	9	0	9	0	1	7	13	4	17
1957	7	1	8	1	1	4	10.5	3.5	14
1958	8	0	9	0	0	5	11.5	2.5	14
1959	2	0	2	1	1	2	3.5	2.5	6
1960	12	3	13	5	2	6	17	9	26
1961	15	1	16	1	0	3	17.5	2.5	20
1962	17	0	17	3	3	6	21.5	7.5	29
1963	12	4	12	4	0	4	14	6	20
1964	15	3	16	7	4	7	21.5	12.5	34
1965	9	2	10	3	0	6	13	6	19
1966	14	2	17	2	0	5	19.5	4.5	24
1967	12	1	15	3	2	5	18.5	6.5	25
1968	21	2	22	3	3	7	27	8	35
1969	5	1	5	2	0	9	9.5	6.5	16
1970	13	1	19	2	0	4	21	4	25
1971	8	6	13	8	1	7	17	12	29
1972	24	0	27	1	3	10	33.5	7.5	41
1973	7	0	9	0	0	2	10	1	11
Totals	247	38	276	65	33	116	350.5	139.5	490

The total sample covers 490 classified cases plus 10 non-classified (difficult) cases.

Table A3 Types of innovations by sectors and by time (abbreviations and definitions: see Figure 1)

Industrial sectors and corresponding SIC numbers (in brackets)

Years	Mining (103, 121, 131)	Building Materials (324,325, 327,329)	Iron & Steel (331, 339)	Non-ferrous Metals (333-335)	Chemicals (281,282-287,289)	Petroleum Refining (291)
1953	I	I	I	I, TM	3 x TM, 6 x MED, P, 4 x IP, I	TM
1954			2 x I	I		
1955	I	I	2 x I		P, I	
1956	I				TM	
1957		TM	I			
1958	I			I	I	
1959				TM	TM	
1960		2 x I			MED, P, I	
1961		I			I, TM	
1962		I	2 x I		I, MED	I
1963			I, 2 x TM		2 x P, D, I	
1964		TM	I		MED, 2 x TM, 2 x P	I
1965		2 x I, TM			2 x TM, P	
1966			4 x I	2 x TM	TM	
1967			I		SI	2 x I
1968	TD, I	P	4 x I		TM, I, D	
1969		TM		TM, I	TM, I, MED	
1970		TM			3 x I, 3 x TM	I
1971		I		I		
1972	I	I, SI	TM		3 x I, 4 x TM	2 x I
1973			I, SI		TM	I

Table A3 Types of innovations by sectors and by time (continued)

Years	Rubber & Asbestos (301)	Saw-Mill & Timber Processing (242)	Woodworking, Cellulose, Paperboard (261)	Steel Construction (344)	Cars (371)	Machinery Construction (351,353-359)	Ship-building (373)	Aircraft Construction (372,376)
1953	2 x TM, P		MED		4xP, I	2 x I, P	2 x I	2 x IP
1954	IP				IP	2 x I		D
1955		TM			P	2 x I		TM
1956				TD	TD	2 x I	I	
1957						IP, 2 x I		I
1958	TD					TM, 2 x I	I	TD
1959								
1960					I	IP, I		P
1961					I,P	2 x I		I
1962	TM					TD, TM, 6 x I		IP
1963					P	3 x I		
1964				I		IP, TD, 2 x I		
1965						3 x I		TM, I
1966					P	4 x I	I	P
1967						IP, 4 x I		P
1968						2 x IP, 5 x I	2 x I	I
1969								D
1970					P, SI	3 x I		I
1971	P				P, IP	2 x I		2 x SI, TM
1972	IP			IP	2xI, IP	TD, 4 x I		TD
1973					I			

Table A3 Types of innovations by sectors and by time (continued)

Years	Electrical Equipment (360,362, 365-7,369)	Precision Engineering & Optics (381-4,386)	Hardware & Metal Goods (341,342, 347,349)	Fine Ceramic & Glass (321)	Wood Manu- facture (243,249)	Musical Instruments, Toys, Jewellery etc. (393)	Paper & Board Manufacture & Printing (262)
1953	2 x I, IP	MED, IP		TM, IP		I	
1954	2 x P	TD					I
1955	TM		2 x I				
1956	2 x TD	I	TM		TM		
1957	TD		I		TM		
1958		SI					
1959	TD	I		IP			
1960	I, IP, 2 x TD	MED, TD, SI					
1961	2 x I, SI, TD						
1962	2 x TD, I	I, MED					
1963	I, TD	I, P					I
1964	I, TM, P	2 x MED, I, SI	I, TD	TM			
1965	P, TD	SI, MED	TM				
1966	3xI,D,SI,D	I, SI	TM, I				
1967	2xTD,2xI,SI	2xTD, MED, I					
1968	SI, MED, TD	2 x I, TM	D	TM			
1969	5xTD, D	I, P	TM				
1970	3xI, SI	MED, 2xSI					
1971	I,2xP,TD	2xP,3xSI,TD,MED I					I
1972	3xI, 2xTM	2xI,TM,SI, MED I		TD			
1973	I	I, SI					

Table A3 Types of innovations by sectors and by time (continued)

Years	Plastics Manufacturing (307)	Leather & Shoes (311)	Textiles & Clothing (222, 228, 231)	Food, Tobacco & Beverages (203, 206, 208, 209)	Non-manufacturing sectors (services, trade etc.) (154, 161, 162, 173, 374, 401, 422, 432, 442, 452, 461, 481, 483, 489, 491, 494, 495, 508, 632, 731, 739, 769, 806, 891, 892, 919, 957, 962)
1953					3 x I
1954	I		TM	IP	2 x TD, TM, I
1955	TD, I				3 x I
1956					4 x I
1957			IP		I, SI, TD
1958			P		3 x I, TM
1959					I, MED, D
1960					4 x I, 2 x TD, P, TM
1961	TM				7 x I
1962	I			2 x IP	3 x I, MED, D, TM
1963					5 x I, TD
1964	3 x IP, I				5 x I, MED
1965	I				2 x I
1966					SI
1967	TM			TD	2 x I, IP, SI, MED
1968	P, TM				5 x I, IP, TM
1969					2 x I
1970		I			I, 2 x SI
1971					2 x I, MED
1972	3 x TM		I		2 x I
1973					3 x I, SI, TD

Table A4
Types of innovations in 8 innovative growth industries

	Pure I	Pure P	I+SI	P+MED	IP	T=T_D+T_M	I+SI+½T +½IP	P+MED+½T +½IP	Totals
1953	4	6	4	13	8	6	11	20	31
1954	1	2	1	2	2	1	2.5	3.5	6
1955	2	2	2	2	0	3	3.5	3.5	7
1956	1	0	1	0	0	4	3	2	5
1957	1	0	1	0	0	1	1.5	0.5	2
1958	1	0	2	0	0	2	3	1	4
1959	1	0	1	0	0	2	2	1	3
1960	3	2	4	4	1	3	6	6	12
1961	5	1	6	1	0	3	7.5	2.5	10
1962	5	0	5	2	1	3	7	4	11
1963	3	4	3	4	0	1	3.5	4.5	8
1964	4	3	5	6	3	3	8	9	17
1965	2	2	3	3	0	4	5	5	10
1966	4	2	6	2	0	1	6.5	2.5	9
1967	5	1	7	2	0	5	9.5	4.5	14
1968	4	1	5	2	0	4	7	4	11
1969	2	1	2	2	0	6	5	5	10
1970	8	1	12	2	0	3	13.5	3.5	17
1971	1	6	6	7	1	6	9.5	10.5	20
1972	12	0	13	1	2	8	18	6	24
1973	4	0	5	0	0	1	5.5	0.5	6
Totals	73	34	94	55	18	70	138	99	237

The 8 industries cover 237 classified cases,
plus 7 non-classified (difficult) cases.

References

W.J. Abernathy and J.M. Utterback (1975) A Dynamic Model of Process and Product Innovation, 'Omega', vol.3, no.6, pp.639ff.

R.W. Coombs (1983) 'Long Waves and Labour Process Change', paper for Conference on Long Waves, Maison des Sciences de l'Homme, Paris, March.

C. Freeman, J.K. Fuller and A. Young (1963) The Plastics Industry: A Comparative Study of Research and Innovation, 'National Institute Economic Review', November, no.26.

C. Freeman, J. Clark and L. Soete (1982) 'Unemployment and Technical Innovation: A Study of Long Waves and Economic Development', London, Frances Pinter.

C. Freeman, L. Soete and J. Townsend (1982) 'Fluctuations in the Numbers of Product and Process Innovations 1920-1980', paper for the OECD Workshop on Patent and Innovation Statistics, Paris, June 28-30.

Gellman Research Associates (1976) 'Indicators of International Trends in Technological Innovation', Jenkintown, Pa., report prepared for the National Science Foundation (PB-263738).

D. Gordon, R. Edwards, and M. Reich (1982) 'Segmented Work, Divided Workers', Cambridge University Press.

A.H. Kleinknecht (1981a) Observations on Schumpeterian Swarming of Innovations, 'Futures', vol.13, no.4.

A.H. Kleinknecht (1981b) 'Prosperity, Crisis and Innovation Patterns', Research Memorandum 1981-23, Frije Universiteit Amsterdam, Fakulteit der Ekonomische Wetenschappen.

E. Mandel (1975) 'Late Capitalism', London, New Left Books.

G. Mensch (1975) 'Das technologische Patt', Frankfurt, Umschau.

G. Mensch (1977) Indizien fur eine Innovationslucke, 'Wirtschaftsdienst' VII, pp.347ff.

J.J. van Duijn (1983) 'The Long Wave in Economic Life', London, George Allen & Unwin.

6 Applying the biological evolution metaphor to technological innovation

Ugo L. Businaro

The biological evolution metaphor

The use of expressions taken from the language of biology when dealing with innovation is certainly not infrequent. Natural evolution was more explicitly used as a metaphor by Nelson and Winter, first in dealing with economy and then to develop a model for technological innovation.(1)

The natural evolution metaphor might look quite simple to a non-specialist, who will mainly refer to a simplified Darwinian two-stage process: mutation (invention) and selection (innovation).

A closer analysis shows a much more complex and conflicting situation in the realm of biologists, which might be very stimulating when compared with the variegated analysis attempted by students of the technological innovation process. To discover striking parallels among the two sets of literature, the reader is referred to an unpublished paper by the author.(2)

When dealing with biological evolution, one should refer to three different points of view: that of the paleontologist, that of the biologist, and that of the molecular biologist.

The first point of view, for which we will refer to Grasse,(3) deals with discovering and explaining the phyletic evolution of the biological world. The second point of view, for which we refer to Dobzhanski,(4) deals with the evolution of single species through the study of populations. The third approach, for which we could refer to Monod,(5) tries to understand the basic principle of biological evolution at the biochemical level.

The different points of view emphasize different aspects of biological evolution and this might explain why, one hundred years after Darwin, the students of biological evolution still debate very aggressively among themselves, defending different theories. The paleontologist tends to emphasize the finality of evolution, while the molecular biologist focuses on mutation and selection (survival of the fittest).

The search for a unifying theory has characterized the last half century of evolution research. We will refer here to the 'synthetic theory'(6) as representing the metaphor for the innovation process.

The basic ingredients of the metaphor are the following:

(a) a process for generating ideas or inventions characterized by creativity and chance,
(b) a 'storage container' where invention can be cumulated,
(c) a 'duct with an on-off valve' that connects the invention storage to the selection device,
(d) a 'selection machine' to test the inventions, accepting only those which are fitted to the 'environment'.

With respect to a simplistic chance-and-necessity theory, there are two major differences. First, the ultimate fate of an invention is not decided at the moment when it appears, so that more favourable future conditions could govern the ultimate selection. Second, the state of the valve (open or closed) is governed by a complex feedback mechanism that depends on history. This is due to the architectural constraints imposed by the 'solution' already developed (the biological individual or the product and related manufacturing process) that the invention aims at modifying. These two ingredients in the model (storage and go/non-go valve) are responsible for some of the most interesting features that appear to be characteristics both of biological evolution and technological innovation: the existence of a 'preferred path' of evolution (chreods or 'necessary path', according to Waddington)(7) and of different speeds of evolution in different times and conditions.

The existence of different eras of accelerated and large innovations in the phyletic change, followed by a period of exploitation of the 'basic' invention, is best expressed by Grasse.(3)

In Drosophila populations,(14) genetic changes appear to diffuse, filling an ecological niche, with the characteristic features of the logistic curve. Mutation in the DNA cannot be fruitful unless in the cytoplasm contains the right enzyme to 'read' the new words formed in the genetic code. When, following a completely different chain of events at organ and gland level, new enzymes are produced, then a stored unused mutation in the DNA could produce sudden important changes.(5)

The Russian doll model

Koestler presents his theory of the 'holon'(8) in a way which reminds one of the sets of Russian dolls which fit inside each other. He uses the word 'holon' to mean a unitary and complex system. At a certain level of observation of the world one sees it as a holon. Going down to more detailed levels of analysis, one sees a holon at each level of analysis. The same happens if one goes up to the level of aggregation. The resulting global image of the world is therefore that of an infinite set of holons, each included in all larger ones, and including all smaller ones.

The dream of breaking down our view of the world into elementary components (the reductionist approach) seems to be impracticable.

Referring to the case of biological evolution, it is interesting to note that the view put forward by Koestler seems to apply - at a certain level of aggregation (that of the paleontologist, or biologist, or molecular biologist) one can look at the observed data as having a system behaviour. The fact that each holon shows the same basic dynamic characteristics might simply be a 'topological' characteristic of open systems (refer to Le Moigne(9) for an analysis of the 'evolution' of an open system).

One should now ask to what extent this model of Russian dolls also applies to the case of technological innovation, a process which encompasses very different levels of 'world view': from fundamental science to applied research, to development and industrialization.

To force a correspondence with the three levels of aggregation presented for the biological evolution case, we will also refer to three levels of aggregation when enquiring about technological innovation:

(a) the epistemologist's point of view, such as that of Kuhn,(10)
(b) the diffusionist's point of view, as typified recently by the work of Marchetti,(11)
(c) the macro-economic point of view, as pursued by Freeman(12) and by Abernathy and Utterback(13), studying long-term innovation change in industrial sectors.

The theories put forward to explain the basic facts analysed from each of the three vantage points are consistent with the basic metaphor described above.

Popper's 'conjectures and refutations' theory (14) (focusing on the holon of the individual scientist) parallels that of simple chance-and-necessity Darwinism. Kuhn(10) with his normal science emphasizes the 'architectural' constraints of history, which seems to force the search along preferred paths. We will refer below to the contribution of Feyerabend.(15)

The success of the 'substitution analysis',(11) when applied to so many different sets of data, can be interpreted with the same model used by biologists in studying the diffusion of a best-fitted population in an ecological niche.

The three-stage model(13) for interpreting the technological change of an industrial sector (from a stage of flux when product innovation prevails, searching for a successful design, to a maturity phase where incremental process innovation prevails) looks very similar to Grasse's(3) model of philetic evolution.

The catastrophe model

The dynamic of a closed system is governed by the increase in entropy. Open systems, on the contrary, in their input-output interaction with the environment, tend to increase the varieties of their configurations and their complexity to the point where, through a crisis or a catastrophe, a new structure of the system is produced.(16)

The simplistic model of an open system which we have described above, using the biological evolution metaphor, can explain the dynamics of the system between one crisis and the other. The selection mechanism is the basis for assuring that the system keeps itself in equilibrium with the environment after it has adapted to it. The effect of the reservoir is that of strategy: in other words, to be prepared for changes in the environment. The biologist, to underline these two abilities of a biological system, distinguishes a 'normalizing selection' (tactic ability to retain suitability to the environment, notwithstanding the push to change coming from mutations) from a 'directional or balancing selection' (strategic ability to maintain adaptation to the changing environment, taking advantage of the stored mutations).

This model, though, does not explain sudden 'catastrophic' changes. The literature of biological evolution can help in pursuing the metaphor's application to technological innovation.

At the paleontologist's level of aggregation, phyletic evolution has presented four or five big, revolutionary changes. At each one of these crises a 'mother form' appears from which various phyla develop. Some of the phyla go through a specialization process and evolution might come to an end, either with a species maintaining a static equilibrium with the environment or with the disappearance of the species.(17) Other phyla maintain more archaic characters, develop less specialization and new attributes apparently of no use for the species itself (e.g. mammalian characteristics in reptiles), up to the moment when they converge in a new 'mother form'.

Increase of complexity of body organization and of the level of psychism characterizes the changes from one stage of phyletic evolution to another. 'Progress' in nature therefore seems to have a precise chracterization: the ability to manage increasingly complex structures by means of an increasing ability to process information (higher psychism).(3)

At the biologist's level of aggregation, the appearance of a new species (speciation) is the equivalent of crisis changes. How can speciation be explained? Are new species generated by a continuous changing process (anagenesis) or by a sudden change process (cladogenesis)? Three of the mechanisms put forward(18) to explain cladogenesis might be important for our metaphor:

* isolation of the environment (the dumb-bell principle): a population

of a given species living in an environment which separates from the rest (e.g. after a tectonic movement), and stays thereafter completely isolated from other populations, might develop into a new species;
* transplanting pregnant females into a new environment (the founder principle): their progeny could develop into a population from which the selection of stored genetic changes useful in the new environment might result in a new species;
* hybridization among close species, living in overlapping ecological niches.

The first case refers to the effect of building different histories of environment - open systems interaction; the second case to the exploiting of the potential of changes stored in the reservoir in case of a sudden environmental change; and the third case to the interaction among separated open systems to form a new one.

It might be interesting to start using the metaphor to find analogies in epistemology. Feyerabend's(15) contribution to epistemology is that of vindicating the role of the anarchist scientist who, by following unconventional searching paths, produces new ideas (experimental data and theories). To do this he has to **isolate himself from the prevailing environment** of 'normal science' (to behave as an anarchist). The build-up of new ideas, together with the increasing incapacity of normal science to explain accumulated experimental data, might result in a new theory (a new Kuhn's paradigm).

An analogy with the **founder principle** (but also with the importance of carrying forward archaic characteristics in the phyletic evolution to produce a new 'mother form') might be found in the procedures by which a designer operates. In approaching a new problem to be solved, he first of all peruses past solutions to similar or different problems.(19) Making old ideas work in a new set of specifications sometimes results in successful new products. (We refer the reader to discussions among students of the design process(20) to discover references to Popper's 'conjectures and refutations' theory, to the historical cycle theory of philosophers such as Vico and others.) It is also interesting to note how often philosophers, in their search for truth, revisited ancient philosophic theories (e.g. revivals of Eleatic studies by neo-positivists) to find new starting points. With respect to the **hybridization model**, it might be interesting to explore the relevance to the progress of science and technology of the overlapping and convergence of separate disciplines in new interdisciplines (bio-engineering, physio-chemistry, etc.).

Long-term changes in technological innovation

It is a naive remark among electrical engineers to state that complex systems show an oscillating dynamic behaviour because of the complexity of their feedback correlations and the imperfection of their control and instrumentation system. Industrial dynamics(21) have helped to transfer this basic understanding to economic and social systems. When trying to set up a simulation model of a complex system, the need for simplification compels one to distinguish internal and exogenous variables, and to look for the determinants of changes.

Changing the way the subdivision is done in the model brings about different explanations for the cause-effect relationships. The debate is apparent when dealing with the issue of long-term economic changes.(22)

What is here suggested is that any 'complex open system' shows intrinsically a dynamic behaviour that goes through a series of expansions of logistic type followed by catastrophic changes. The interactions among different complex systems (holons?) or, more simply, between a system and the rest of the world (the environment) determine the time scale between one crisis and the other. The Russian doll model and the interactions up and down among the different holons could help us to understand the appearance of micro-and macro-cycles and hypercycles.

It is not our intention to proceed with this train of thought which might soon resemble a philosophical belief. We will simply point out that it is not strange at all, when one concentrates the analysis on the technological system, to discover intrinsic sagging-type dynamics. The analysis of major invention/innovation data is consistent with the metaphor put forward earlier. The storage mechanism, in particular, is quite evident.(23)

That the interactions between different holons are important in setting the time constants of the global system can be perceived in the analysis done by Marchetti.(11)

The scarcity of data on technological inventions and innovations and the ample margin of discretion in their classification leave ample space for debate when trying to define quantitatively the cycling characteristics: Do the waves have a constant period? How many cycles are there from the beginning of the industrial revolution?

To what extent can the metaphor we are proposing help? We have already obtained some hints on how the 'speciation' metaphor can explain changes in the paradigms within the basic science holon. What about the interactions among the different holons internal to the innovation process (basic research, applied research, development)? Giarini and Louberge(24) have proposed an interpretation of the last two hundred years of economic development in terms of two major waves. The first is a technological wave based on the build-up and exploitation of empirical knowledge developed in the eighteenth century, and the second is a wave based on the interaction between technology and science starting in the last part of the nineteenth century.

One could interpet the second wave proposed by Giarini and Louberge as a 'speciation' coming from the hybridization of overlapping species (science and technology). For an heuristic use of our metaphor we propose a less aggregated analysis, at industry sectoral level. At the beginning of the industrial revolution all the industrial sectors were based on empirical knowledge and their development could be seen on the basis of the invention-storage-selection model internal to the holon of development and industrialization. The development of modern mechanics and thermodynamics certainly has helped the development of technological innovation, but in a somewhat indirect way (philosophical positivist attitude, understanding of basic principles and constraint from science, etc.).

The thermo-mechanical sectors have so far developed mainly on the basis of empirical knowledge. Other industrial sectors have developed following a different pattern: they are the so-called science-based industrial sectors (such as electricity and chemistry). Scientific knowledge developed first; the related industrial development can be seen as the exploitation of the niche opened up by scientific discoveries. The relationship between science and technological development has been very strict. The history of development of large companies in the electrical sector, such as General Electric, or in the chemical sector, shows the pushing role of basic and applied research. The fact that these sectors have been based on scientific knowledge had an important effect: their technological development path could be forecast following the scientific discoveries and progress. The similarity with the 'chreods' of the biological metaphor is apparent. The 'ecological niches' of these sectors seem now to be well exploited, and several authors(24) point to the law of diminishing returns of research in such fields.

The situation and the role of research for empirically based sectors is different. Here applied research has had mainly a service role, that of solving problems posed by the practical development of the sector itself.

Direct coupling with basic research has been scarce. What will happen when the empirically based sectors encounter basic science, and scientific knowledge could be used to design a product (predicting its detailed behaviour theoretically, without having to test prototypes empirically)? This might be the beginning of a new 'chreod' of development, a 'speciation' by hybridization of different cultures. There are signs that this is also becoming a reality for very complex products such as vehicle engines.

The 'reservoirs' of basic and of applied research - filled with inventions that have led to innovation in other sectors - will be available for new innovation needs, as soon as the 'selection valves' on the ducts connecting them to the empirically based sectors are open. The limitation of the empirical approach for complex product design has so far permitted only 'good enough' design compromises. The scientific approach will permit real optimization in meeting design specifications and constraints. How far is today's 'good enough' car engine design (which developed through a century of empirically based invention and innovation) from tomorrow's optimal design (based on the possibility, for instance, of predicting on paper the distribution of flow fields and material composition at every point of a combustion chamber)?

While existing scientifically based industrial sectors are approaching the end of the exploitation of their respective chreods, the 'mature' empirically based sectors might be faced with a long-lasting period of new development. The actual situation is more complicated because we are facing not only hybridization between empirical technologic and basic knowledge, but also hybridization with 'horizontal' new technologies such as micorelectronics and informatics.

The coarse picture of technological innovation that emerges above at an aggregated macro-economic level is that of a superimposition of the somewhat separate development of two different classes of industrial sector: one development started from the holon of industrialization, and the other from the holon of basic science. It might be interesting to speculate as to whether there has been a third chreod of development for new sectors starting within the holon of applied research.

Without going too far along this path of speculation, could one not look at the phenomenon of the large applied research projects started during the second world war (radar, nuclear energy) and soon after (space) from this point of view? These projects certainly could have started because of existing consolidated scientific knowledge (successfully passing the selection test in the basic research holon), pointing towards potential practical application (filling the applied-research reservoir of inventions). The establishing of large projects has had the effect of opening the selection valve and providing resources for performing the selection operation itself. It is a 'speciation' coming from a 'pregnant' applied research 'female' transplanted to a new environment (the resources made available by the big projects).

Will these new phyla borne by the 'mother form' of applied research develop into successful new industrial sector-species? The fate of the nuclear industry is still uncertain, but space telecommunications seems to be here to stay.

Time phasing of industrial innovations

In another paper(25), the author has used the biological innovation metaphor to illustrate how the relationships within an enterprise develop among R&D and other functions within a company and whether a rational approach could be following in the subdivision of the financial resources among the different investment needs.

In a large company, manufacturing, design, and applied research can each be considered as a complex open system - separated organizationally and spatially from the others - showing its own 'evolutionary' dynamics, but having complex interactions among themselves and the environment (see Figure 1, where the case of three interacting invention-selection systems is shown, building up the innovation process from basic to applied research to development and industrialization).

It is important to note that the intrinsic time constant is different for the different stages in the innovation chain. For instance, in the case of the car industry, the lifetime of an engine factory is of the order of 20-30 years, that of an assembly line, 10 years. The commercial lifetime of a new car model is around 10 years, with two or three major restylings during the course of the model life. The time needed to design a new car - assuming that all the relevant technical and technological information is readily available - is four to five years. The demonstration that a new concept of engine is feasible, or that new materials can be technically and economically introduced on cars, can range from a few years to 10-20 years. The difference in time constants and the uncertainties intrinsic to R&D projects show how important it is to have the reservoirs of invention in the development and industrialization system well filled at the moment when the decision is taken to proceed with a new product design and new capital equipment investment.

On the other hand, the successful completion of the applied research phase (showing how innovatively to change a component in a product) cannot be transferred before the decision is taken to change the product model or to renew the obsolete manufacturing equipment (the selection valve should be open).

According to our metaphor it has to be expected that the flux of actual innovation (the output at the end of the innovation chain) shows an oscillatory behaviour. Do these oscillations have wave-like characteristics with constant periods? The theory of product life-cycle assumes that products, in a certain product class, have a distinct constant lifetime. In recent years, however, this theory has been much criticized and only in special cases (mature industry, stable market) is a periodicity apparent (see the case of automobiles in the USA in the years 1950-1960). The more common case is that of non-periodic oscillations. The company management is directly responsible for setting the time of each innovation's oscillation because of the decision to start a new product and/or to renew capital investment. The intensity of the innovation in the new product depends less directly on management itself, because of the effect of previous decisions on

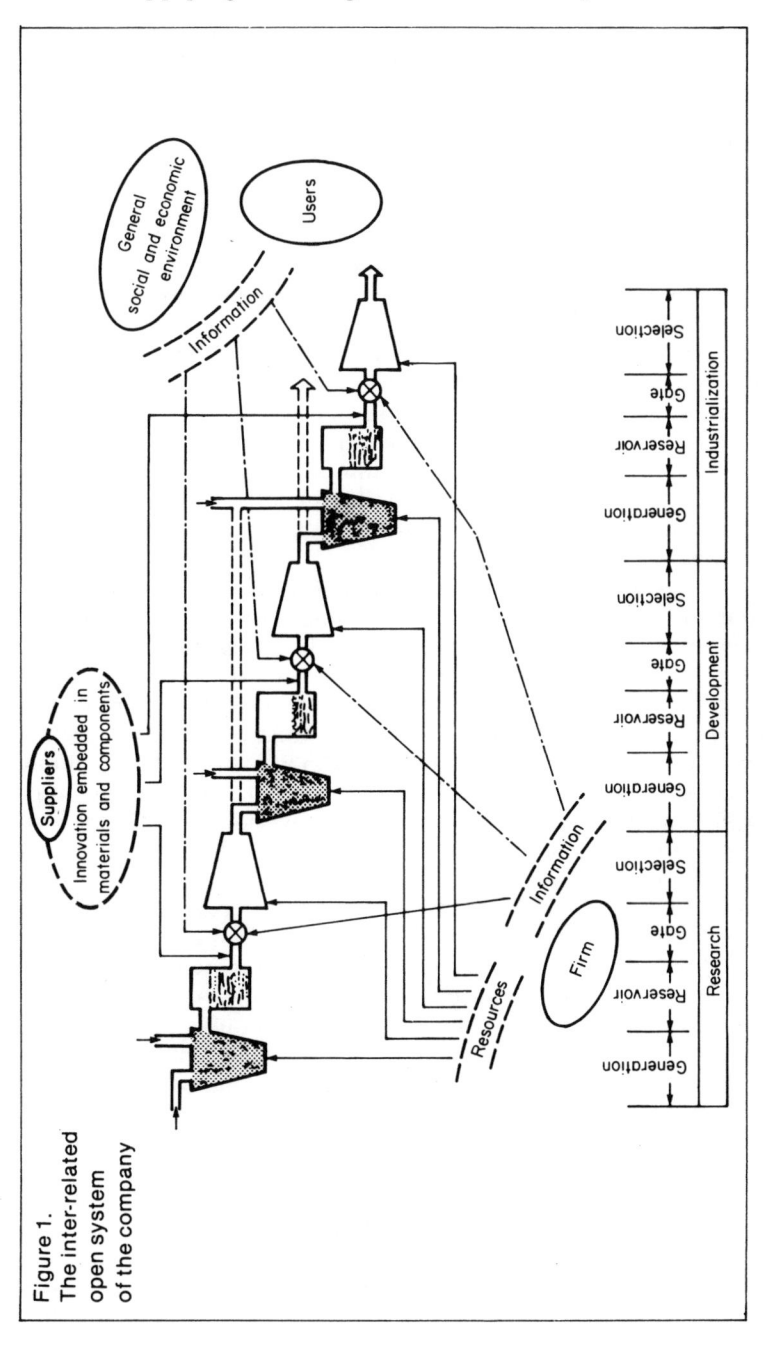

Applying the biological evolution metaphor 113

Figure 1.
The inter-related open system of the company

allocating resources to research projects (opening the selection valves in the R&D system) and the availability of innovation proposals from outside the company (suppliers of materials and equipment, etc.).

When aggregating the individual companies in an industrial sector, and different sectors in the entire productive system, it seems difficult to accept that the aggregation of widely different unphased and unperiodic oscillations will lead to a periodic behaviour. Even the clustering of major innovation seems to be difficult to understand as the result of aggregating the microscopic behaviour of the different actors in the innovation process. The contradiction of the evidence of large innovation changes within an industrial sector (see, for instance, the study by Abernathy of the US car system(26)) might simply be another case showing that in complex system analysis the reductionist approach is not valid (the system shows a global behaviour different from that of the sum of its components).

The metaphor of biological evolution, and especially the reference to the speciation mechanisms, could help in better understanding how a change in the environment could be related to a sudden burst of innovation.

Let us consider that a major economic crisis is the equivalent of an environmental change in natural evolution. What happens at the micro-level of a company when a sustained crisis is on? Each individual company, because of the gloomy market forecast, will try to delay the introduction of new models and the renewal of the manufacturing plants whose planned time of change falls within the period of economic crisis. The old product still has to withstand the competition which is becoming stronger during the market shrinkage. Company management is therefore looking for innovation which can reduce manufacturing cost and somewhat renovate the old product (restyling or face lifting) at as low a cost as possible. The innovation, mainly at component level, should be compatible with the existing product and manufacturing system. Within this constraint, the management is ready to take higher risks, accepting innovations which are not fully proven. In our metaphor jargon, the selection valve is now open. During extended crisis, therefore, the 'selection machine' operates at the component level, and the entire industry structure has an opportunity to learn, directly in the field, how to make the best use of innovation. In other words, the formal barrier between research and industrialization is raised and the two systems proceed together along the learning process.

When the economic crisis is over, the acquired learning about how the innovative changes can be dealt with will diffuse across all the company functions, at the component level. This will be the starting base for innovative jumps, this time at the system level (both product and manufacturing considered as a system). To give an example, the introduction of a microcomputer as a trip computer on cars (mainly motivated as a way to 'face lift' the old product) will familiarize the mechanical world of car engineers with electronics, and this learning will be the basis for a true integration of electronics and mechanics when designing new engines.

As another example, the introduction of automated computerized quality control stations at different points in the car manufacturing lines might be accelerated, in time of crisis; because they are compatible with the existing manufacturing facilities, they increase quality and reduce cost. When a completely new plant is designed, this will permit a change in the entire system philosophy, interconnecting (via computer) all the automated quality control stations to obtain feedback signals which will permit changing the process variables (for instance, changing dies and tools) in order to keep quality within prescribed ranges.

The future of the world of products

When analysing natural evolution, it is difficult to resist the temptation - following the apparent finality of evolution itself - to forecast what will be the next step. Teilhard de Chardin[27] has tried it. The pervasive diffusion of information technology - with the exponential increase in the flow of communications and the building of memories external to our brains - might be seen as a move in the direction of the development of the nous predicted by Teilhard de Chardin. This temptation continues. As an example, one can refer to a recent book by Chauvin,[28] which suggests that increasing data available on the psychic activity of different animals might point in the direction of an evolution towards developing the optimal 'brain'.

The reason for mentioning Teilhard de Chardin and his epigones here is only to apologize - in view of their much more ambitious and far-reaching attempts - for posing the following question and trying to speculate on it: Is it possible on the basis of the 'process' of evolution (not its ends but its means) to forecast the lines of developments of the products made by man? Technological innovation in its broadest sense is the process (the means) used to renew the world of products. It might seem easier here, with respect to the natural evolution case, to accept that the end of progress in the world of products is clear, i.e. to satisfy human needs in increasingly better ways. But is it? How do human needs change because of the appearance of new products?

Fortunately, it is not necessary, in proceeding with our speculatory attempt to answer the above question, to look at the finality of product development. The general process itself by which an open system interacts with the environment seems to condemn it to progress in a certain direction, following a typical pattern. An open system that is born renovated in its structure after a revolutionary change (a catastrophe) exploits the potential of its new structure (fills its 'ecological niche'), increasing the complexity (both internally in the system and in relation to the environment) to a point where an increased ability to deal with it is needed. A revolutionary change might then happen, which would change the system structure and the relationship to the environment.

The Grassé(3) definition of 'progress' for the phylogenetic natural evolution (increasing complexity, controlled by an increased level of psychism) could be translated as a metaphor to the world of products by replacing the word 'psychism' with the word 'information' or 'knowledge'. As a matter of fact, it is not difficult, looking to human history, to trace the continuous increase of knowledge needed to manufacture and/or to use products. The simple case of the vase will illustrate the point. In the prehistoric age, clay and solar heat were used to make pottery. The amount of information needed to learn the process is very small. At the beginning of the historical age people learned to make vases out of glass; the amount of information needed to learn this process is much larger. Think now of a vase made of thermoplastic material, and the amount of information (organic chemistry to develop new materials, manufacturing technology, etc.) that had to be accumulated during the last two centuries to make it possible.

Luckily, one does not have to use all that accumulated information (studying chemistry, thermodynamics, control theory, etc.) to make a plastic vase. The information has been aggregated in easy to use 'packages' (thermoplastic grains, the extrusion press, etc.) so that, because of this high degree of 'order', it is very simple to transfer the needed knowledge to practical manufacture of plastic vases. (Refer to the paper by H.A. Simon, 'The architecture of complexity',(29) concerning the importance of the ability of complex systems to decompose into quasi-independent subsystems.)

The recipe we are suggesting here, to try to forecast 'revolutionary' changes in our world of products (due to 'revolutions' which might be difficult to detect because of being masked by continuous incremental innovation both on products and the manufacturing process), is to look at the increasing degree of complexity in different product sectors (and/or looking from different functions of society), and ask oneself if a new 'order' to deal with such complexity might be needed, and made possible because of a higher-level ability to aggregate and manage information.

For example, it will be suggested here that we look at the problem from three vantage points: that of the materials needed to manufacture products, that of the primary human needs satisfied by the products, and that of the so-called service sectors.

The number of different materials available to build any kind of product has increased exponentially. Even a talented mechanical engineer has difficulty in perceiving the relative advantages/ disadvantages among a host of engineering plastics. Different materials are first adopted and then discarded in new models of a product, to reappear again (often together with a new manufacturing process) in a later version of the same product. The success of applications of new materials in other industrial sectors (e.g. carbon composite in aircraft) inspires designers to look at the possibility of their use in completely different fields (electric motors, cars, etc.). The learning process for optimal use of a new material involves trial and error, and is very lengthy. It took more than 15 years from the first appearance of thermoplastics on car instrument panels to its optimal use today (from the point of view of design style, choice of materials, manufacturing process).

The first wave of the industrial revolution was dominated by steel as the base material for most industrial products. Since then, new materials have been added to the list, with an accelerated pace in the last half century, thanks to developments in chemistry. Are there too many? Could fibre composite materials (together with new process technology developed to increase flexibility of application) become the new base materials? In such a case, the product design itself would be changed, as would today's subdivisions between industrial sectors (primary materials producers, the transformation industry, component and end product manufacturing).

The amount of information needed to design and to manufacture products, making optimum use of composite materials, is certainly greater than that needed to use steel. The problem - in our metaphor for this progress - is to know if such increased information could not be managed by aggregating it, thereby simplifying the entire process of designing, building and using products.

Oversimplifying the actual situation, one could say that products can be grouped according to the primary human needs they satisfy: home, transport, food, etc. These needs do, however, interact to a greater or lesser degree. Product specifications should therefore take account of such interactions. The basic design of a product and the way it satisfies the basic need(s) might have stayed unchanged for several decades, or even centuries (even if the product and its manufacturing technology have undergone a continuous series of innovations).

In the meantime, society (and the way the primary functions interact) might have changed a lot since the first appearance of the product. Is the basic design of the product still the optimal response?

As an example, consider the situation of kitchen-wares. Their number has increased enormously, and often they are left unused in their package in a kitchen which has become increasingly small, especially in dense urban areas. Food is also changing, with an increase in variety from all over the world and in the type of processing (raw, pre-cooked, frozen). The housewife of today is confronted with all the alternatives from grandmother's recipes to frozen TV dinners to microwave cooking. Is it not too complex a situation, offering too much contrast to the changes in other primary needs (home, leisure, etc.)? A re-examination of kitchen-wares with a view towards simplification and a higher level use of available information might be needed. This might result in a 'revolutionary' change in the product world.

As another example, take the interaction of the automobile with urban traffic. One solution proposed by transport planners years ago, but as yet unsuccesful, is to induce a shift in the use of different modes of transport, favouring an increased use of collective public transport. The resistance of car drivers to abandoning their habits, notwithstanding the increased complexity and reduced efficiency of car trips (slow average speed, parking problems, etc.) should point to a new direction which might lead to a 'revolutionary' product change. Here again the answer might lie in the possibility of making higher level use of information management, thanks to the information technology revolution. The new car of the future would then have the ability to interact actively with a computerized traffic control system, not only to optimize the operation of traffic lights, but also to change to an automatic mode of driving on certain properly instrumented lanes. It might be too limited, however, to look only at the changes to the automobile in its interaction with traffic. Its interactions with parking, as well as with other transportation needs, are important, and taken together might lead to a complete redesign of transportation means and infrastructure. In this instance I am not pointing to a 'hard' change, as such would be prohibited by the historical development so far (all the hard investment already made).

A more intelligent use of 'soft' technology at the design stage (e.g. suiting car length to parking spaces and rail transporters) and an improved use of the existing infrastructure 'intelligent' traffic lights, instrumented lanes, etc.) might do a lot. In the end, a major revolutionary change (a simpler way to manage a dense and highly mobile society) might result. In changing primate to man, nature did not rebuild the body, but simply added a small cortex to the brain!

The third case comes from the service sectors. An increasing share of the active population works in the tertiary sectors, and for several years social scientists have claimed that we are shifting to a post-industrial society. This does not mean, however, that fewer 'hard' products will be produced, replaced by 'soft' products or services. A recent study by Gershuny(30) is illuminating in this respect. While occupation in the service sector is increasing, this merely means that the hard products we buy have a higher content of service activities in their added value. While industries are buying more and more industrial services, the reverse seems to be the case for personal services. If one looks at the way personal services are performed, one can detect a trend of increasing complexity and reduced efficiency (school, health, social services). Gershuny suggests that new 'hard' products are emerging which, together with the availability of an increased amount of free time due to a reduction in working hours, are increasing the possibility of replacing bought personal services with do-it-yourself work. Examples range from substituting the safety razor for the services of the barber, to using taped lessons instead of teachers (as in the case of The Open University), to the possibility of using family computer terminals and special detectors for preliminary medical check-ups instead of a visit to the doctor.

Enough has been said in these last years about the effect of the information revolution. What is suggested here is that its revolutionary effects, still to be seen, will come about because of the increased complexity of our society's 'open system' due to the success of the latest wave of developments (the affluent society, increased social security and social equality, world-wide person-to-person communication and interaction). This increased complexity is leading to a decrease in efficiency, first apparent in social services. No matter how popular a politically conservative approach might become (dreaming of a return to the 'good old days' when society was less affluent and less complicated), the process of progress in an open system will find a solution by using the information technology available at a higher level of intelligence, simplifying the use of complex knowledge by 'packaging' it differently, and restoring the role of the individual, possibly starting in the service sector itself.

References

1. R.R. Nelson, S.G. Winter, In search of a useful theory of innovation, 'Research Policy', 6, 1977, p.36.
2. U.L. Businaro, 'Comparing natural evolution and technological innovation', FIAT Research Center Internal Report, 1982.
3. P.P. Grassé, 'L'évolution du vivant, Matériaux pour une nouvelle théorie transformiste', Paris, Ed. Albin Michel, 1973.
4. F. Dobzhansky, 'Genetics of the evolutionary process', New York, Columbia University Press, 1970.
5. J. Monod, 'Le hasard et la nécessité', Paris, Ed. du Seuil, 1970.
6. E. Mayr, 'Animal species and evolution', Cambridge, Harvard University Press, 1963.
7. C.H. Waddington, Stabilization in Systems: Chreods and epigenetic landscapes, 'Futures', April 1977.
8. A. Koestler, 'Janus - Esquisse d'un système' (French translation), Paris, Calmann-Levy, 1979.
9. J.L. LeMoigne, 'La théorie du système général', Paris, Presses Universitaires de France, 1977.
10. T.S. Kuhn, 'The structure of scientific revolution', Chicago, The University Press, 1962.
11. C. Marchetti, Invention et innovation: les cycles revisités, 'Futuribles', Mars 1982, p.43.

12. C. Freeman, 'The economics of industrial innovation', London, Penguin, 1974.
13. J.M. Utterback, W.J. Abernathy, A dynamic model of process and product innovation, 'Omega', vol.3, no.6, 1975, p.639.
14. K.R. Popper, 'The logic of scientific discovery', 1934.
15. P.K. Feyerabend, 'Against method: Outline of an anarchist theory of knowledge', NLB, 1975.
16. J. Prigogine, J. Stengers, 'La Nouvelle Alliance, Metamorphose de la Science', Paris, Gallimard, 1979.
17. J.L. Heim, '700,000 siecles d'histoire humaine', Paris, Eyrolles, 1979.
18. M.J. White, Speciation: is it a real problem?, 'Scientia', 1981, p.455.
19. J.C. Jones, 'Design methods: seeds of human futures', New York, J. Wiley & Sons, 1980.
20. R. Jacques, J.A. Powel (eds), 'Design : Science : Method', London, Westbury House, 1981.
21. J. Forrester, 'Industrial Dynamics', New York, McGraw-Hill, 1966.
22. Special issues on Long Waves Futures, October 1981 and June 1982.
23. See Figure 18 in A.K. Graham and P.M. Singe, A long wave hypothesis of innovation, 'Technological Forecasting and Social Change', 17, 1980, p.283.
24. O. Giarini, H. Loubergé, 'La civilisation technicienne à la dérive', Paris, Dernon, 1979.
25. U.L. Businaro, 'R&D investments and business cycles, Development of the corporate strategy', Paris, EIRMA Annual Conference, 1982.
26. W.J. Abernathy, 'The productivity dilemma, Road block to innovation in the automobile industry', Baltimore, Johns Hopkins University Press, 1975.
27. P. Teilhard de Chardin, 'Le phenomene humain', Paris, Ed. du Seuil, 1955.
28. R. et B. Chauvin, 'Le modele animal', Paris, Hachette, 1982.
29. H.A. Simon, 'The sciences of the artificial', Cambridge, MIT Press, 1969.
30. J.C. Gershuny, Social Innovation: Change in the mode of provision of services, 'Futures', December 1982, p.426.

7 Design trajectories for airplanes and automobiles during the past fifty years

J.P. Gardiner

During the past fifty years, major changes in the means of personal transport have occurred. In Western developed countries, for distances less than 300 miles, cars have become the dominant means, and for more than 300 miles, aircraft have become the dominant means of personal transport. Railways are still important for commuting and for some long distance travel in Europe. The most startling decline has been the virtual extinction of long distance sea travel. There are now only a few liners for leisure cruise trips, and a small but growing business in short distance passenger and vehicle ferrying operations. This paper will attempt to compare and contrast the design trajectories of the personal means of automotive and air transport. As interesting as they might be, we will not be able to do more than lightly touch upon related design trajectories for military and cargo transport vehicles.

What are design trajectories for airplanes and automobiles?

In discussions of technological direction, for any generalization there are exceptions that prove the rule and exceptions that break the rule. Whether or not one sees a generalization as being proved or refuted very much depends on one's point of view. To philosophers, this would seem to be a very subjectivist and eclectic position, and altogether not logically very neat and tidy. But to designers, working in a technological environment trying to produce better goods and services, such philosophical nit- picking is just a waste of time. In design terms, the world frequently seems very contradictory and the things that are designed for use in it often have just as contradictory requirements and specifications. Automobiles are designed essentially to be powered wheeled vehicles for use on land. And yet, we now worry about aerodynamic shape, knowing full well that they will never fly (or at most only be temporarily airborne in exceptional cases). Equally, aeroplanes are designed essentially to be powered flying vehicles for use in the air. When it came to designing a flying machine as big as a Boeing 747, the problem was not whether it would fly, but rather, once you had it flying, could you then successfully land at speed something that weighed more than three-quarters of a million pounds. The crucial design problem for the Boeing 747 (and its nearly as heavy predecessor, the Galaxy C-5A) was to make them into high-speed, heavy-wheeled land vehicles. This was achieved with their innovative multi-wheeled landing gear designs, which involved eighteen or twenty wheels. For working designers, seemingly contradictory requirements are just part of the business, and if that worries philosophers, then they are welcome to it.

This excursion into the philosophy and the problem of technological design contradictions has a point when we come to trying to describe design trajectories. To paraphrase our opening remark:

> For any design trajectory there are exceptions that prove the trend and one-offs that violate the trend.

Since both airplanes and cars are used as technological means of human transport, we could - but will not - sketch out some sort of technological transport trend. It might start with a man in a carriage, next a man in an early car or aeroplane, then a man in a subsonic or supersonic aircraft, and finally, a man in a space rocket. While this might be of interest to some people, such as those in social forecasting, it is of limited interest to designers because the underlying technological changes are very different. Even in this paper, the design trajectories for automobiles and aeroplanes will be kept separate. What will be argued is that during the past fifty years there have been some interesting parallels and some divergencies due to forces outside of those in control in these two industries. At the end, there is a more speculative discussion to explore the possibility that the current automotive design trajectory may be curiously very like that for commercial aircraft some fifty years earlier.

In trying to describe a design trajectory, it is easiest to start with the current state of the art and then work backward into history or forward into the future. By making this move, a design trajectory can be characterized by a dynamically evolving change in the state of the art at a succession of points in time. Practically, this is more easily said than done, because we now have to find a way of describing differing states of the art at different times.

All patents normally have a section describing the then state of the art and how the invention 'improves on' or 'differs from' it. With a lot of research, one could possibly work out a history of changes in the state of the art of various technologies, and then go on to interpolate some sort of design trajectory. Having read thousands of patents, I find there are all sorts of objections to this approach, not least the problem of what subset or collection of patents is used. Keyword title descriptions and patent classification codes could quickly get one into nightmare-sized problems or yards of computer printout. While patents range from just two pages to over a thousand, most are ten, twenty or thirty pages in length. Taking an average of twenty pages, and my personal scanning rate of one page per minute, then a subset of 10,000 patents would take 3,300 hours, or nearly two working years (with some allowance made for holidays to try to remain sane!). Most other literature-based surveys would have similar sorts of problems. The other more important point is that while patents are of some importance to the automobile industry, the aviation industry is somewhat more variable. Aeroengines are heavily patented, but airframes are not. Airframes are covered by Registered Designs in many countries and by Design Patents in the USA; unfortunately, these are limited disclosures with nothing said about the then state of the art.

Within most industries, the typical way of describing the current state of the art is to point and say 'Well, that's about it ' where the 'it' may be a house, a vacuum cleaner, an electricity power station, or a coal-mining machine. At the beginning of Ford's Fiesta (Bobcat) development programme, the state of the art to meet or exceed was the Fiat 127, and just before the Fiesta's launch in 1976, the state-of-the-art design was re-targetted on the then new Volkswagen Golf. This operational way of picking out specific examples of the current state-of-the-art design is not completely without problems, but it is the way working designers do it, and it captures the dynamic element for design trajectories as it is re-targetted at different points in time. While different people can choose different examples as representing a current state of the art, it is this system that we will adopt because then we can get started, and because things are made explicit and open to argument (and possibly change) if someone has a better suggestion.

The other strength of using real design objects - be it a car or an aircraft - to help describe a design trajectory is that whole bundles of dimensions are implicitly taken up. Most economic or technological trends have just a few dimensions, such as maximum speed vs. time, or vehicle weight vs. time, or cost per mile vs. load and time. Unfortunately, two- or three-dimensional analyses of technological or economic trends present only part of the picture. An aircraft or car is a bundle of characteristics and dimensions with major compromises made among them to try to balance requirements which are often nearly contradictory. Each decade seems to demand new compromises, and if one is just looking at two- or three-dimensional economic or technological trends, then what may have been a new smooth line becomes kinked by something like the noise and pollution regulations of the early seventies. A design trajectory built upon state-of-the-art real cars or aeroplanes contains many economic and technological dimensions and their compromises suitable to that given time and place.

A few simplifications

Having said all this, ideally one should go on to try to produce some design trajectories for cars and aircraft. Before we can start, some simplifications will have to be introduced because there are thousands of examples to draw on, each with different characteristics. For example, even if we were looking only at the basic shape and configuration of aircraft, and nothing else, there is an almost bewildering array of examples (see Fig.1).

In terms of commercial importance and overall personal use, we are only going to look at medium- and long-range non-military aircraft, and cars in the light and medium classes. Short-range aircraft and extra-small or luxury or sports-type cars need their own design trajectories, which involve too much work to try to set out here. The next simplification has to do with what are to be chosen as the particular examples of automotive or aviation state-of-the-art designs. The history of automotive and aviation technologies has some consensus, but the biggest difficulty is trying to disentangle the experts' and one's own nationalistic pride and biases. As a Canadian who has worked and lived for almost equal amounts of time, during my adult life, in Europe and North America, I hope that in what follows I have been able to strike some sort of balance in assessing the design contributions from these two main areas of economic activity.

Since the 1930s, American firms have led and dominated much of the medium- and long-range commercial aircraft business, with only Rolls Royce and Airbus as European-based organizations making much in the way of important contributions. Equally, in the 1920s and 1930s, American automobile firms produced new and important designs that were commercially very important. It is arguable that in the post-war period, European car makers (who had been much more disrupted by the war) were eventually the ones who turned out to be the leaders in terms of new state-of-the-art automobile designs. During most of the fifties and sixties, American car designs simply became more ornamented, decorated and heavier. Many American design examples ended up being twice as heavy as comparable European cars, having a 4/5 seating arrangement. American cars gained with a somewhat larger trunk/boot, but lost substantially when it came to such things as fuel economy (and, some would say, handling). In the seventies, a 'downsizing' programme was instituted by the major American car makers. This continuing programme has led some commentators (Jones, 1982) to see American designs converging with those of the Europeans, though still with some distance to go.

If one allows the preceding assumptions and simplifications, then the aviation design trajectory is based mainly on American state-of-

Design trajectories 125

AIRCRAFT CONFIGURATIONS:
1 Santos Dumont 14 Bis
2 Pou du Ciel
3 Lee-Richards circular monoplane
4 Miles M.39 Libellula
5 Blohm und Voss BV 141
6 Fairey Battle
7 Chance Vought Pancake XF7U
8 Blériot XI
9 Arsenal Delanne
10 Dunne tailless
11 Focke-Wulf Ente
12 Rockwell B-1
13 BAC Lightning
14 Junkers Ju 287
15 Handley Page HP 115
16 Saab Viggen
17 Convair XF-92A
18 Puffin
19 Lockheed P38 Lightning
20 Northrop XP-79
21 Chance Vought Cutlass
22 North American X-15
23 Concorde

Figure 1. Source: Mondey, 1977, p.90.

the-art examples, while the automobile design trajectory (particularly post-war) is based mainly on European state-of-the-art examples. As a simple temporal series, the two design trajectories are presented in Figure 2. Given these two design trajectory time lines, then decade by decade comparisons can be made. These are summarised in Table 1. Following a discussion of these major design trajectory changes, there are a few final observations and speculations about the general pattern of automotive and aviation developments.

In Appendices 1 and 2, some of the main technical parameters are given for the state-of-the-art examples used in describing the design trajectories for aircraft and automobiles. Some of these state-of-the-art examples are further analysed in a second paper entitled 'Robust and Lean Designs' in these conference proceedings.

Some observations on automotive and aviation design trajectories during the past fifty years(1)

The thirties

During a period of depression when innovative activity might also be depressed, automotive and aviation designers switched away from structurally based timber construction to all-steel and duraluminium monocoque structures respectively. Equally, while V8 and radial engine designs had been around at least from the time of the first world war, it became feasible to mass produce them economically and with power levels two to three times higher than it had been possible to achieve previously. Perhaps somewhat more surprising, the mating of more powerful engines and new all-metal monocoque bodies was dramatically successful in both cases. Ford's V8 model and Douglas's DC-3 rapidly gained major market shares in depressed economic conditions and quickly established new states-of-the-art for automotive and aviation design.

The forties

Aside from some spin-off type production developments in large press lines and transfer lines, most automotive designs and examples entered a period of suspended animation from 1938 to 1948.
 Military needs gave a tremendous stimulus to aircraft design. Many new four-engined bombers and transports appeared, mostly with new air-cooled radial engines. These radial designs were mass produced as interchangeable power units, and (with supercharging or turbocharging) led to the doubling or trebling of cruising altitudes, payloads and ranges. Military requirements also led to much more inherent redundancy and consequently reliability of designs, and to the beginning of what we now call air traffic control systems, for the

Design trajectories 127

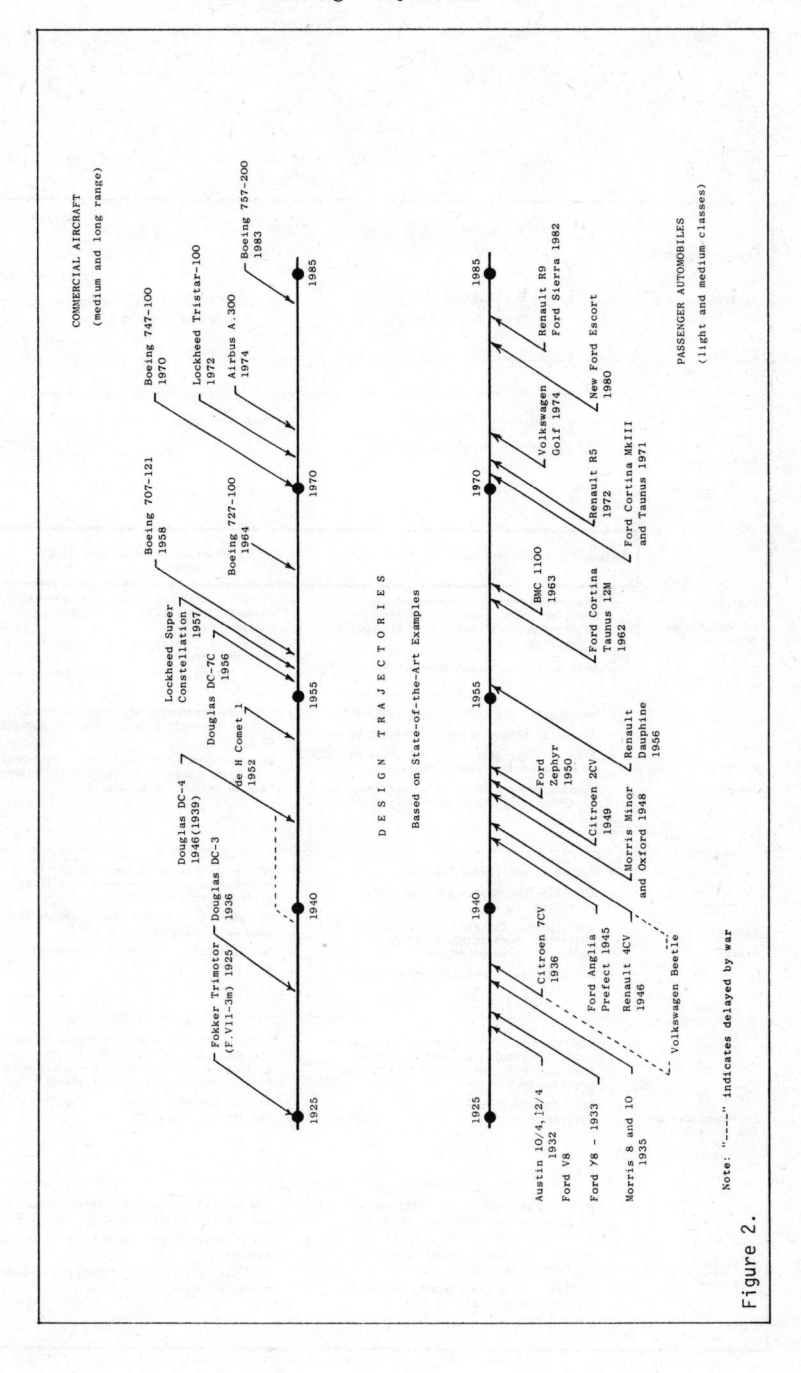

Figure 2.

Table 1

ECONOMIC CLIMATE by decade	MAJOR CHANGES IN DESIGN TRAJECTORIES	
	● AUTOMOBILES (Light and Medium Classes)	● COMMERCIAL AIRCRAFT (Medium and Long Range)
THIRTIES Depression	+ All steel enclosed monocoque bodies + "New" inexpensive V.8 engines - Straight 8's begin to decline	+ All (dur)aluminium monocoque structures + "New" very powerful radial engines
FORTIES War and Postwar Recovery	Automobile production switched to military needs + Bigger and heavier presses and new transfer lines developed for first military production and then automobile production ± Most models: continuation of pre-war series	Aircraft production switched to military needs + Large four-engined, long range aircraft converted to civilian use. Some are pressurized + Jet engine development begun with a military priority - Large flying boats come to an unexpected dead end
FIFTIES Growth	+ New chasis-less unitary body construction + Automatic transmissions for higher hp cars + Front or rear powertrain and drive designs (some based on pre-war designs) - Turbine powered units tried but found to be technically and economically lacking	± First turbocompound and then turboprop powered aircraft have their brief moments of glory + New intercontinental and trans-continental turbo jets appear. Airlines believe they have to re-equip with jets
SIXTIES Growth	+ Higher volume production + Reliable low, medium and high speed operation - Small producers taken over or go out of business; more standardisation of models by large producers + New materials - mainly plastics	+ Special purpose airplanes for short, medium, and long range work at different passenger volume levels + Reliable, all weather short, medium and long range flights + New materials - plastics, fibers and titanium
SEVENTIES Environmental and Fuel Crisis	+ Engines re-designed to meet new noise and pollution regulations and to improve fuel economies while often maintaining previous power levels - More radical Wankel design does not succeed	+ Engines designed and re-designed to meet new noise and pollution regulations and to improve fuel economies while often maintaining previous thrust levels - Radical Concorde design does not set a new design trajectory
EIGHTIES Recession	+ Many new evolved designs with better performance and operating economies e.g. Volkswagen Formel E Golf + New microelectronic based power train management systems - Consumers have trouble affording new more economic models	+ New evolved design with better performance and operating economies e.g. Boeing 757 - New microelectronic based avionics systems - Airlines have trouble affording new more economic models

safe take-off, handling and landing of large numbers of aircraft. Heavy, four-engined, wheeled bombers also stimulated the development and building of very long reinforced concrete runways, which at the time were innovations in civil engineering. Immediately after the war, the conversion of many bombers and transports to civilian use and the availability of long land runways virtually blocked any further development of large transcontinental and intercontinental flying boats as the main direction for a commercial aviation design trajectory. This was a complete surprise to most designers and operators who had grown up in the thirties. In the USA, a worry that there might be a shortage of aluminium, and a somewhat nostalgic hope for large flying boats, resulted in what turned out to be one last parting shot: the 700-seat Hughes Hercules, more popularly known as the 'Spruce Goose', of 1947. Even later in the UK, there was the Saunders-Roe (Saro) Princess (Figure 3).

The fifties

This was a period of economic growth and the cold war. Both factors influenced automotive and aviation design trajectories. Pent-up post-war demands and general economic growth led to the extension of earlier design trajectories and at the same time to the appearance of new, more radical design trajectories. Overlaying this situation, the existence of the cold war and the consequent additional military demands on the aviation sector led to its complete transformation into the 'jet age'. On the automotive side, there was a partial transition to a new, chassis-less, unitary body construction, and to powertrains driving the wheels directly opposite them, whether they were at the front or the rear (e.g. 2CV or VW 'Beetle'). This design transition is now more or less complete for the light car class, but currently in the medium car class there is still a mixture of front-wheel drive and traditional (e.g. the R9 and the Sierra). In the 1950s, the transformation of automotive designs into a wholly new radical trajectory, based on unit bodies, front-wheel drive, and turbines working through continuously variable transmissions, was tried but never successfully put into production.(2)

For commercial builders and operators, the conversion to the wholly new, more radical jet age design trajectory was initially a bit unexpected, and then enthusiastically embraced by both sides because of the mutual expectations of increased business in a period of general growth. The reason the jet age was a bit unexpected was threefold:

1. Initially, turbojets were essentially a by-product of the military, and somewhat 'uncivilized'.
2. Existing radial designs were turbo-compounded, with large power units dramatically increasing from about 2500 to around 3500 hp.(3)

130   Design, innovation and long cycles

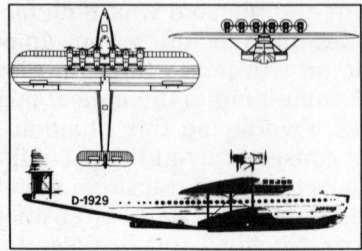

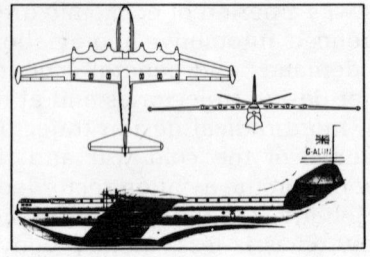

Figure 3. Source: Mondey, 1977, p.302.

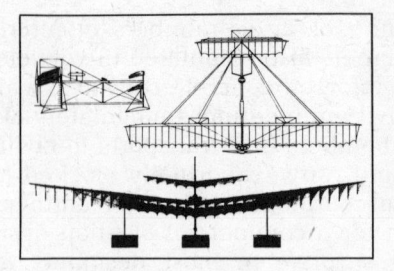

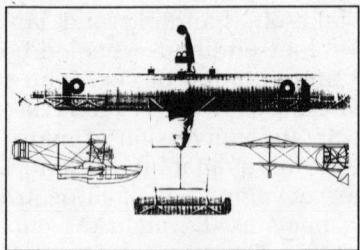

ABOVE: *Henri Fabre's Hydravion was the first successful floatplane. Built in 1910, it was powered by one 50hp seven-cylinder Gnome rotary engine, driving a pusher propeller, giving a maximum speed of 90km/h (55mph).*
RIGHT: *The Short Sunderland, which first flew with the RAF in 1937. It carried seven to ten Browning and Vickers machine guns and a bomb load of 700kg (2,000lb).*
BELOW: *The Dornier DO X, the world's largest aircraft when it was built in 1929. It had a maximum speed of 215km/h (134mph), and a range of 1,700km (1,056 miles). It could carry up to 150 passengers.*

ABOVE: *The Navy Curtiss NC-4, the first aircraft to cross the Atlantic, did so in several stages. It was powered by four 400hp Liberty V-12 engines, two pusher, two tractor, giving a maximum speed of 145km/h (90mph) and a range of 2,360km (1,470 miles).*
BELOW: *The Saunders-Roe (Saro) Princess, a very large flying boat intended as a high capacity transatlantic transport aircraft. It first flew in August 1952, powered by ten Bristol Proteus turboprops, eight of which were coupled in pairs to contra-rotating propellers. These engines gave a cruising speed of 580km/h (360mph) and a range of 8,480km (5,270 miles).*

The author wishes to thank Octopus Books
for permission to use the illustrations
from D. Mondey (ed.), "The International
Encyclopedia of Aviation", 1977.

3. The then new turboprops were seen as offering the best of both worlds.

Nevertheless, first the Comet and then, even more so, the Boeing 707 of 1958 rapidly convinced builders and operators that the jet age had arrived. (Comet and 707 designs are discussed in the paper 'Robust and Lean Designs'.)

The sixties and early seventies

For commercial aircraft and mass market automobile builders, the sixties was a period of general growth and improved specific performance levels, only to be followed by one of externally imposed constraints. These new constraints were, of course, due to environmental regulations concerning noise and pollution, and to the economics of substantially increased fuel charges.

The general growth in demands for aviation and automotive transportation in the 1960s led to higher and better performance levels in car and jet design. The natural extension for some were new design trajectories based on large 300-400+ hp American cars weighing over two tons, and on the proposed Mach 2 Concorde Supersonic Transport-type aircraft. Events of the 1970s proved them not to be new design trajectories, but interesting sidelines.

(At this point, we should perhaps stop and reflect a moment. In reviewing the above history, it is all too easy to be clever with the aid of hindsight. Design trajectories are not uniquely defined and determined into the future. Flying boats of the thirties and large American cars and supersonic transport designs of the sixties were in their respective periods believed to be the emerging design trajectories. They were to be stopped short by external factors - namely, war in the first case and the environment and fuel crises in the second. Generally speaking, things (and particularly cars and aircraft) have got bigger and better over the years, but the problem is that they do not keep getting bigger and better in the same ways that they did previously. Perhaps this is only another way of restating our earlier remark that almost inevitably two- or three-dimensional analyses work for a while, but there comes a point when the trend line or curve gets kinked. Consumers have a multitude of non-constant requirements. Technological artefacts such as cars and aircraft have in turn bundles of dimensions with changing weights and combinations through time. Design trajectories as described here are not really a long-term forecasting tool. What automotive and aviation design trajectories do describe, by reference to state-of-the-art examples, are past, present and probably near future (i.e. five to seven years ahead), and the relative balance and mix of different dimensions. The future has always been uncertain, but design trajectories do suggest what the next step into the future might be.)

The seventies marks something of an economic watershed. Prior to this time, it was often primarily a question of whether something could be achieved in technological terms (e.g. speed or acceleration), with the economics tending to follow along in a more secondary role. From the mid-seventies onwards, economics has dominated the technical dimension in the sense that economics sets the primary design brief and then the technological factors follow on in a more secondary way. By economics, we mean here the internal operating costs of a car or aircraft (e.g. fuel, maintenance, depreciation) and the external costs (e.g. noise pollution, automotive or aerospace industrial employment).

The introduction of the Concorde and the Boeing 747 nicely represent in their respective ways the case of an earlier time having technological priorities and a later time having economic priorities. See Figure 4. In this figure we have also plotted a point for the new Boeing 757. The 757 is a replacement for the 727, and the primary specification of the design brief was that overall operating costs would have to be cheaper than those of the existing 727. To this end, most recent reports suggest that in 1983 the 757 is 10% cheaper to operate than a 727.

(This was worked back to the 1975 costs of Figure 4 so that the 757 could be plotted on that chart. Incidentally, in doing this we can illustrate what we mean by trends being kinked at some point, because if the Boeing 757 had been plotted as at 1983 instead of at the date of introduction of the Boeing 727, then we would have had to introduce a nearly horizontal line to join up with the 1983 point (even if everything was still in 1975 costs). This process is not only true for new aircraft, but increasingly so for new cars, where generally the new model has to have at least as good overall operating economics in real terms as its predecessor, and then at the same time be technologically better on several other points, such as acceleration or passenger comfort.)

The late seventies and early eighties

Our recent recession has increasingly brought into focus the economics of internal and external operating costs. On the second, which is beyond the scope of this paper, there are government policies and programmes on both sides of the Atlantic involved with aerospace and automotive investment and employment considerations, and with environmental and fuel policy priorities. What we can look at are the internal all-in operating costs of aircraft and automobile users.

If we take a new standard medium-sized family saloon with an average of two of the seats occupied, and a new medium-range jet with standard fleet operating characteristics, then perhaps somewhat

Design trajectories

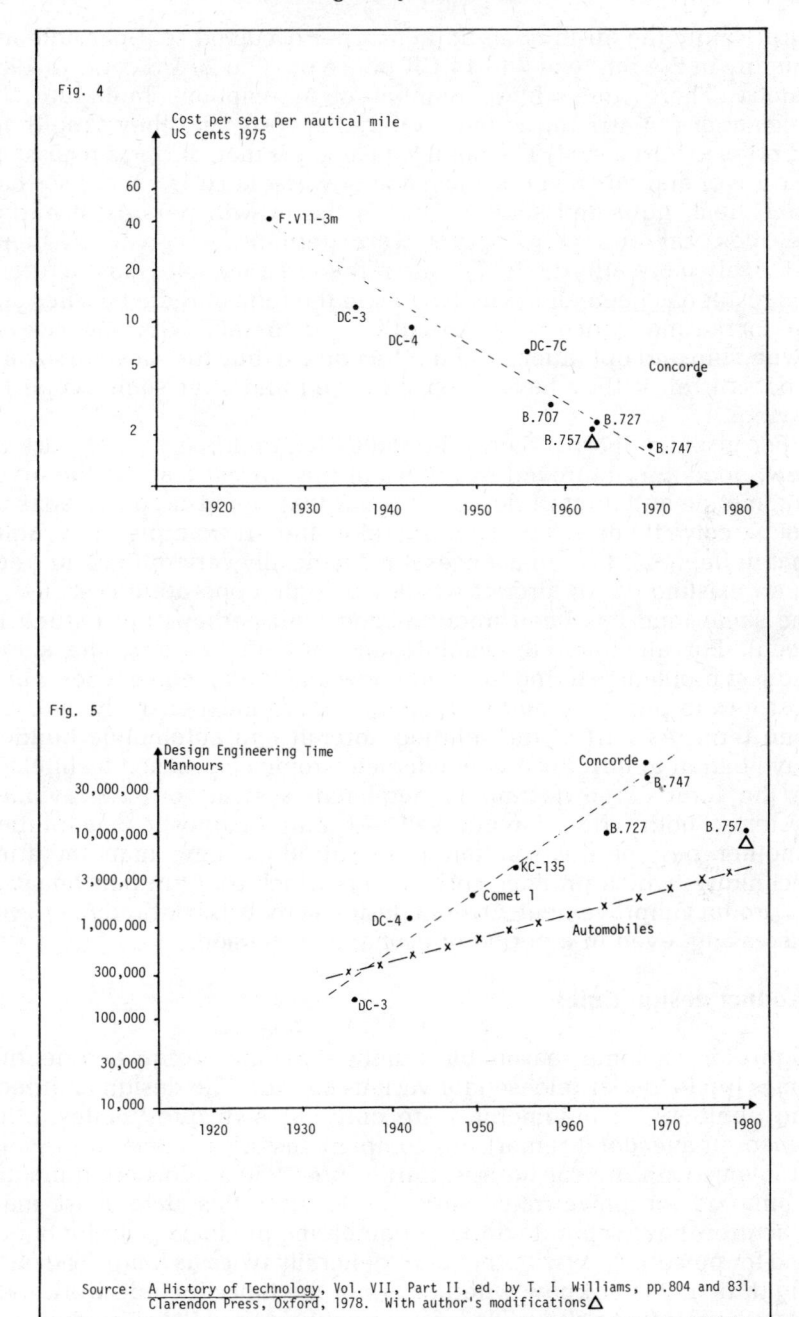

Fig. 4

Cost per seat per nautical mile
US cents 1975

- F.VII-3m
- DC-3
- DC-4
- DC-7C
- B.707
- B.727
- B.757 △
- B.747
- Concorde

Fig. 5

Design Engineering Time
Manhours

- DC-3
- DC-4
- Comet 1
- KC-135
- B.727
- Concorde
- B.747
- B.757 △
- Automobiles

Source: A History of Technology, Vol. VII, Part II, ed. by T. I. Williams, pp.804 and 831. Clarendon Press, Oxford, 1978. With author's modifications △

surprisingly the all-in operating costs per occupied seat per mile are roughly in the range of 7 to 14 UK pence or 10 to 20 US cents in 1983 values. There are a huge number of assumptions built into the calculations in arriving at these operating costs, and they should not be pressed too hard.(4) The point to be made is that, during a recession, car travel and air travel are personally variable costs - more so than food, heat, light and shelter. During the growth periods of earlier decades, car and plane travel were perhaps perceived as being relatively more affordable. For aircraft and automobile manfacturers, the result has been all too noticeable in the form of reduced sales and an increasing concern by would-be purchasers with the cost of ownership and operation, and a reluctance to buy the most up to date models even if they have been improved and offer some scope for savings.

For instance, Ford's Sierra, Renault's R9, and Boeing's 757 are all new, good and improved examples of the current state of the art in automobile and aircraft design. It is not that would-be purchasers do not perceive them as being good state-of-the-art examples; it is rather that in depressed circumstances it is financially very difficult to trade in an existing car or aircraft which has higher operating costs but at the same time has been underutilized (and perhaps not earned its keep). For all major automobile and aircraft builders, the single biggest problem is trying to design new and improved vehicles which cost less to purchase and also operate more inexpensively over the long term. As part of the solution, aircraft and automobile builders have begun to introduce new microelectronic chip-based technology in the form of powertrain management systems or new avionics systems, both often having self-test and diagnostic capabilities. Another part of the solution is to introduce new manufacturing techniques which produce cost savings which could in part be spent on product improvement. Overall design activity is continuing or even increasing, even in a period of economic recession.

Product design times

Figure 5 has some reasonable numbers for the design engineering times (up to 'paper release') for various aircraft. The design engineering numbers for automobiles are only our own guestimates, often based on anecdotal remarks in company histories. There are several problems with the car figures. Earlier pre-1950 models often include manhours for powertrain work, while after this date most manufacturers have separate design engineering product cycles for bodies and for powertrains (with the latter generally twice as long). The other big difficulty is to know whether or not the times include work done (i) with prototypes and mules; (ii) dissecting competitors' models; and (iii) arriving at production engineering costings for the new designs.

Given the logarithmic scale of Figure 5, errors in arriving at precise numbers tend to be compressed when looking at overall trends.

Aviation and automobile design engineering times have both been generally upward, with the former increasing more than the latter. The divergence would be even greater if design engineering times were worked out on a per-unit-of-production basis. When many civilian aircraft had a military equivalent, much of the high design engineering time and costs could be set against military development programs and production runs. Since the seventies, there has not been a need for equivalent military units, and aircraft production and design engineering efforts are being rationalized. An estimated number for the Boeing 757 has been added to Figure 5 and it shows a marked divergence from the previous long-term trend. Boeing has rationalized its design engineering and production for the 757 and 767 so that they have shared development programs, and final production will see something like 42% or more commonality between the 757 and 767. Shared development programmes and production commonality have long been a part of the automotive business and have helped keep design engineering overheads (per unit of production) well down.

Having said all this about trends and divergencies of quantitative design engineering efforts, perhaps a more important remark should be made about the nature of qualitative efforts. From the seventies onwards, computer aided design (CAD) has been increasingly used in automotive and aviation development work. If CAD had not been used, cost and manhours would have been substantially more and probably prohibitive. CAD has got rid of a lot of routine draughting/copying work and allowed some designers to be more creatively employed. The bigger impact in design terms for cars and aeroplanes has been the ability to model alternatives which would have been too expensive to take through to the prototype or production stage. Having done this, it is frequently possible to arrive at better compromises and composite designs. This feature is a very important element in creating a robust design, of which more will be said in another paper.

A few final observations and speculations

Similar and dissimilar design trajectories

Automotive and aviation design trajectories show similar underlying directions and shifts, partly due to internal technological developments and partly due to external, mainly economic factors.

The biggest divergence between the two was due to the design requirements and funding levels of the military for the aviation sector and not for the automotive sector. The result was that the commercial aviation business moved on to a more radical technology in the fifties:

namely, jet engines and subsonic aircraft. At the same time, the automotive business experimented with the more radical technology of turbines and continuously variable transmissions. This costly line of development was terminated because, in the past, there was not a large special interest user such as the armed forces who was prepared to spend the large sums required. In the seventies, and now particularly in the eighties, the interest of the armed forces in military derivatives or variants of commercial aircraft has declined as their requirements have focused on other parts of aerospace technology (e.g. various guided missile systems).

Design trajectories in the eighties(5)

Without military interests, commercial aviation has returned to an environment very much like that of the thirties. Aviation and automotive firms are both faced with a situation where the only course open to them seems to be to come up with new designs and innovations just to survive. As in the thirties, intense efforts are being made to improve the performance of powertrain units, with a particular eye to improving both short- and long-term economics for car and commercial aircraft owners. And by improved economics (in real terms), we mean either improved technical specification at the old cost or substantially improved technical specification at a somewhat increased new cost. In design terms, this means at least matching the existing state of the art at a slightly reduced cost in order to leave a margin for slightly improved specifications or at least move the state of the art on a step which will justify to the prospective owner an increase in initial real costs, but with an eye to reducing long-term overall costs. Renault's R9, Ford's Sierra and Boeing's 757 are all recent examples of these approaches. Overall, the eighties is likely to be a decade of steady evolutionary improvement and technical change with a great deal of fairly standard design and redesign work. While the developments may be 'fairly standard', the state of the art will be moved on and any aircraft or car builder who fails to keep up will probably go out of business in what can only be increasingly competitive markets.

None of the major firms in the automotive or aviation business seems to be willing to move back to much earlier states of the art and costs, to produce something like a contemporary version of the post-war Morris Minor or DC-4 which have perhaps half the currently accepted performance levels for cars and aircraft respectively. In France, there are very small firms such as Ligier (who are better known for their Formula I racing cars) which are now producing a few thousand Micro cars per year, with room for just two people, a three and a half hp 49 cc engine, and modest performance in towns only. The French Micro cars are interesting because they are way outside the design brief specification level of the smallest small car in the Small Car Class of any of the major automobile producers. In the eighties, Micros could be the true personal town car that electric cars of the seventies were hoped to be, before electricity charges increased and optimistic forecasts of battery performance were not reached.

A new aircraft builder and an automotive design divergence in the eighties

Having argued in the previous section (with the exception of an additional Micro car class for the eighties) that we are in for a decade of incremental design changes and innovations, most professional academics would stop there. There is good reason to do this, because the state-of-the-art examples, while showing the obvious effects of age, generally hold true for five to seven years in the future of both the automotive and aviation design trajectories. With examples such as Ford's Sierra, Renault's R9 and Boeing's 757, I should be safe until the late eighties - and shut up. But I won't. Here I have a feeling of deja-vu rather than professional scholarship that urges me to go on.

1. The Japanese position: In the automotive and aviation design trajectories, there are no Japanese state-of-the-art examples. Why not? In cars, Nissan/Datsun licensed pre-war Austin state-of-the-art examples, and post-war, by the seventies, began to emerge as a major force in the world automotive business. Since then, several Japanese firms have been generally quite successful(6) with the very interesting strategy of producing state-of-the-art examples that are two to three years out of date, but with advanced state-of-the-art production techniques which ensure a very attractive final cost. (Ongoing research at SPRU by Jones and Gardiner hopes to document this more precisely in the future.) In aircraft in the thirties, the Japanese bought one of the first low-winged all-metal duraluminium monocoque aeroplanes (i.e. General Aviation's Clarke G.A.43, sold to Mitsui Bussan in 1934). Also, one of two DC-4 prototypes was sold to them in 1939. Mitsubishi brought out the G.3M commercial version of a long-range medium bomber which made an impressive and fast world tour in late 1939. Post-war, in 1957, the Japanese Ministry of International Trade and Industry set up a Transport Aircraft Development Association which produced the successful two-engined turboprop YS-11 for Nippon Airways in 1965. Given the Japanese ability in technological intelligence, and capability of turning intelligence into successful commercial production, one really wonders if they will not appear in the major commercial aircraft business in the 1980s.

2. New car bodies and powertrains in the eighties: The aircraft business of the thirties, even more than the automotive business, was transformed by new power units and body-building techniques. The same ingredients seem to be in the mix for automobiles in the 1980s.

For aircraft there were super/turbo charged engines and variable pitch propellers. Already we have commercial production of turbocharged automotive engines (some with fuel injection) with microelectronic management systems.(7) What we are still lacking is the equivalent of a variable pitch propeller or continuously variable transmissions. On the body side, we are part way there with HLSA (High Strength Low Alloys) steels and plastics. There were forming problems with duraluminium and currently deep drawing problems with HLSA steels which probably can be solved. While we are not likely to see all-plastic cars, we could see HLSA steel structures and plastic skins and interiors. Again, most interiors and bumpers are now plastic based - the only thing lacking is plastic skins.

Ford has a very interesting patent (see Figure 6) which describes a new glass-fibre reinforced plastic sheet material which can be pressed, fixed, painted or sealed in almost the same way as conventional steel sheet material, and can be used in conjunction with a basic steel structure. The DC-3 emerged midway through the thirties. A car with an HSLA structure plus plastic skins and a turbocharged injected engine, plus continuously variable transmission (with a microelectronic powertrain management system), may yet emerge in the mid-1980s. Late in the eighties we could go on to turbines if there are good continuously variable transmissions.

Just as in the thirties there appeared to be many reasons for being pessimistic, while in reality there were somewhat hidden reasons to be more optimistic, much the same can be said of the automotive and aviation businesses and their design trajectories in the 1980s.

(12) **UK Patent Application** (19) **GB** (11) **2 057 350 A**

(21) Application No 7929582
(22) Date of filing 24 Aug 1979
(43) Application published 1 Apr 1981
(51) INT CL³ B32B 5/26 // 5/28 27/04 27/36
(52) Domestic classification B5N 0526 0528 2704 2736
(56) Documents cited
GB 1366091
GB 1347176
GB 1342147
(58) Field of search B5N
(71) Applicants
Ford Motor Company Limited,
Eagle Way,
Brentwood,
Essex.
(72) Inventors
Emmit W. Archer,
Francis Derek Gentle.
(74) Agents
R.W. Drakeford,
15/448,
Ford Motor Company Limited,
Research & Engineering Centre,
Laindon,
Basildon,
Essex.

(54) Reinforced sheet plastics material

(57) Polyethylene terephthalate (PET) sheet material incorporating fibrous reinforcement, eg. 20 to 60% glass fibre, is made by holding superimposed layers of reinforcement, 3, and molten PET, 1, in contact under pressure and then cooling the layers to below the glass transition temperature of PET at a rate sufficient to avoid crystallization of the PET. The resulting sheet material contains PET in its amorphous phase and can be moulded to form articles such as motor vehicle body panels by heating the sheet in a mould to a temperature above its glass transition temperature but below its melting point and holding the sheet at that temperature until the PET undergoes crystallization.

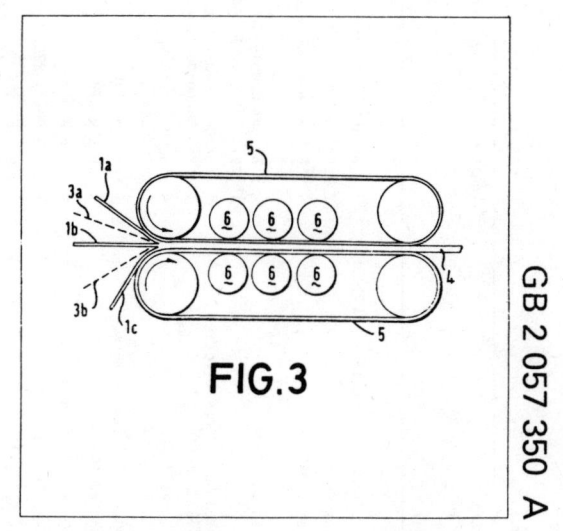

FIG.3

The drawing(s) originally filed was, were informal and the print here reproduced is taken from a later filed formal copy.

Figure 6. Reproduced by permission of the Controller of Her Majesty's Stationery Office.

APPENDIX 1 State-of-the-Art Examples for Commercial Aircraft Design Trajectories

Aircraft	Engines	Entry into Service	Typical No. of Seats (Econ.)	Typical Maximum Range (km)	Typical Cruise Speed, V (km/hr)	Gross Weight W (Tonnes)	sfc (kg/kg/hr)
Fokker Trimotor (actually F.VII-3m)	3 x Wright Whirlwind	1925	8	650	160	5.2	0.18
Douglas DC-3	2 x P & W Twin Wasp	1936	21	1,450	270	11.3	0.28
Douglas DC-4E	4 x P & W Twin Hornets	1939 (1946)	44	3,400	322	30.2	
Douglas DC-7C	4 x Wright Double Cyclone 18 cyl. turbo-compound	1956	80	7,500	500	63	0.39
Lockheed Super Constellation (L.1649 "Starliner")	4 x Wright Double Cyclone 18 cyl. turbo-compound	1957	90	8,000	470	72	0.36
de H Comet I (HS)	4 x de H Ghost 50 (centrifugal)	1952	44	3,200	790	45	1.17
Boeing 707-121 (early mark)	4 x P & W JT3C	1958	165	5,300	850	112	0.90
Boeing 727-100	3 x P & W JT8D-1	1964	131	3,200	970	77	
Boeing 747-100 (Jumbo)	4 x P & W JT9D	1970	490	8,000	900	335	0.65
Lockheed Tristar-100	3 x R-R RB211-22B	1972	330	4,300	885	194	0.64
Airbus A.300-B2	2 x GE CF6-50A	1974	280	3,200	840	137	0.63
Boeing 757-200	2 x R-R RB211-535E4	1983	185	5,600	860	100	

Source: The Technology and Economics of Air Transport in its Next Phase, K. G. Wilkinson, Aeronautical Journal, March 1976, p.127. With some authors modifications.

APPENDIX 2

State-of-the-Art Examples for Automobile Design Trajectories

Automobile Make and Model	Production Dates	First or Basic Engine Size c.c.	Engine Horse-power at ... rpm	Miles per Gallon	Wheelbase	Weight cwt	Comments
Austin 10/4	1932-36 ... 47	1125	30 @ 3800	34	7'9"	15½	
Austin 12/4	1932-36	1535		27	8'10"	20	
Ford V.8	1932-38	3622	65 @ 3400	16	8'10"	23	"Streamline" Model
Ford Y.8	1933	939	22 @ 4000	32	7'6"	13¾	
Morris 8	1935-39 ... 48	918		45	7'6"		
Morris 10	1934 ... 48	1378		25	8'6"		
Citroën 7 CV	1936 ... 57	1600					"Traction Avant" and Unitary Construction
Ford Anglia E04A	1945-48	933	23 @ 4000	36	7'6"	14¾	
Ford Prefect E93A	1945-48	1172	30 @ 4000	33	7'10"	16	
Renault 4CV	1946-61	760	19 @	45			Rear Engine
Volkswagen "Beetle"	1946 to date	1131	25 @				Côntinuously Modified
Morris Minor	1948-52 ... 72	919	27.5 @ 4400		7'2"	14¾	
Morris Oxford	1948-56	1476	41 @ 4200		8'1"	19¾	
Citroën 2CV	1949	375	9 @				F.w.d. Unitary Construction
Ford Zephyr	1950-56 ...	2262	68 @ 4000	23.7	8'11"	22	
Renault Dauphine	1956-	845	30 @				Rear Engine
Ford Cortina Mk.I & II	1962-70	1198	48.5 @ 4800	30	8'2"	15½	
Ford Taunus 12M	1962-70	1200					F.w.d. and V-4 Engines
BMC 1100	1962-74	1098	48 @ 5100	33	7'9"	16¼	F.w.d. and Hydro-lastic Sus.
Ford Cortina Mk.III	1970-77	1298	57 @ 5500	25	8'6"	19¼	Same as Ford Taunus
Renault R5	1972	845	37 @ 4400	42	7'10"	16½	
Volkswagen Golf	1974	1093	50 @ 6000	35	7'10"	14¾	Jetta Variant in 1980
New Ford Escort	1980	1117	55 @ 5700	36	7'10"	15	
Ford Sierra	1982	1294	60 @ 6000	30	8'7"	19½	
Renault R9	1982	1108	41 @	44	8'2"	16¾	
Below not shown on Fig. 2							
Ford Model A	1928-33	3285	40 @ 2200	19.4	8'8"	22¼	
Ford Anglia and Prefect 100's	1953-59	1172/ 2262	36/68	30/24	7'3"	14½	

"..." indicates continued later in this form or nearly so.

F.w.d. indicates front wheeled drive.

Notes

1. In the account which follows there are several implicit but important threads in the story of design trajectories which will be dealt with in the paper on 'Robust and Lean Designs'.
2. In some as yet unpublished work by the author, some recent patents survey work suggests that this more radical design trajectory is again being considered, but with the addition of microelectronic management systems.
3. Turbocompounded 18-cylinder radials gave the DC-7Cs and Super Constellations true intercontinental performance, with 7,500-8,000 km ranges.
4. These figures are the result of the author's estimate based on all-in costs derived from organizations such as the Consumer's Association 'Motoring Which ' and personal communications from the aircraft industry.
5. Some as yet unpublished and incomplete patents survey work underpins the conclusions in this section.
6. Wankel engines seem to have been one of the few failures.
7. W. Downs of Ricardos has argued that what we need are design families of engines, with a basic model and up/down related versions, and four heads for regular fuel, indirect injection, direct injection and mixed (alcohol) fuel. This results in twelve possible variants, of which perhaps half would be in production at any one time.

8 Robust and lean designs with state-of-the-art automotive and aircraft examples

J.P. Gardiner

Throughout the earlier discussion of automotive and aviation design trajectories, there was reference to state-of-the-art examples to help describe them. At this point we need to look a bit more at the nature of the particular state-of-the-art examples that were chosen. In doing so, we have in part some explanation of their specific qualities that made them good state-of-the-art examples. Quite simply, most of the good state-of-the-art examples have ROBUST designs.

There were a few state-of-the-art examples that in the end failed to live up to their initial promise of establishing a new direction for a design trajectory and these, typically, were based on LEAN designs.

In the progress of any active technology, there are always new ideas and sometimes a few prototypes that embody them. Most designers and engineers will express some interest in them, but some scepticism as well, because new ideas mean a divergence from that which has been found to be tried and true. For us, a robust design is one that brings together several new divergent lines of development to form a new 'composite' design, which is then internally adjusted to form a new 'consolidated' design, which is then further developed as a variety of 'stretched' design. Lean designs fail at one, two, or more often all three of these stages.

This preceding robust design model can be set in the context of a broader design-innovation model (Rothwell and Gardiner, 1983). In terms of this model, most aviation and automotive robust designs are instances of re-design between the innovation and re-innovation stages. For an outline of the more general model, see Figure 2, and for an example of a robust design and the invention-innovation stage, see Appendix 1.

Much of this story is not new, and the first two phases in the development of a robust design are the same as for the 'dominant designs' of Abernathy and Utterback (1976) in their well-known model of industrial innovation. Most dominant designs are robust designs, and vice versa, but not always. Abernathy and Utterback's Model T is a dominant design, but not a robust design, because it failed to be stretched as a product in various up/re/de-rated versions. (The man-

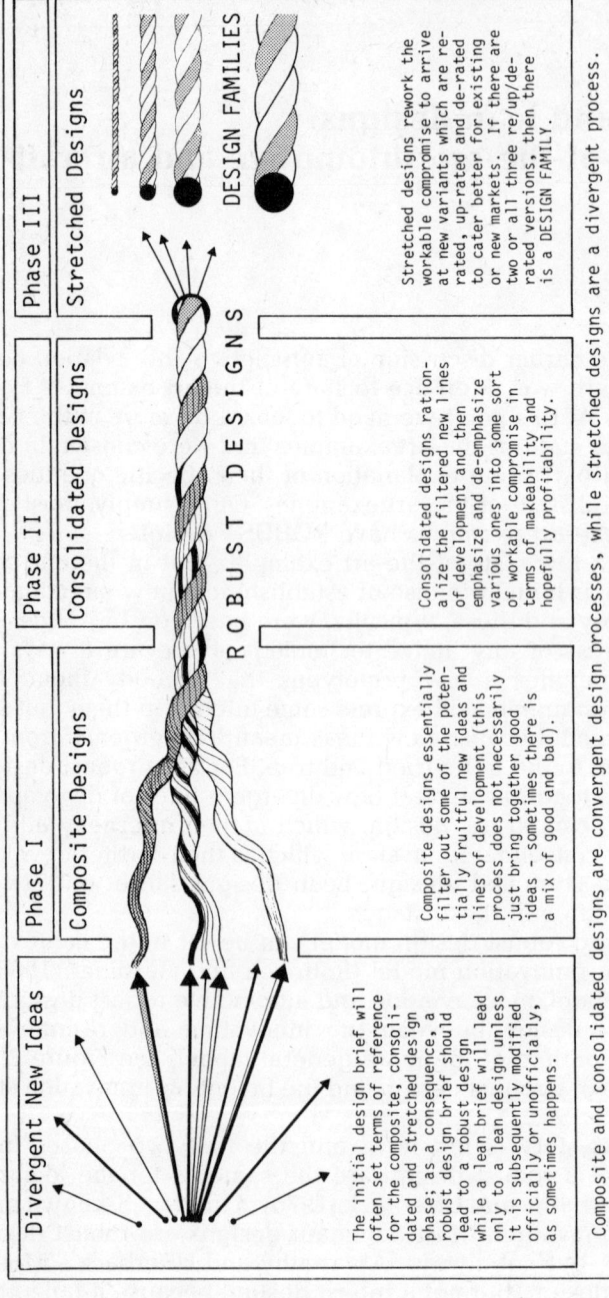

Figure 1.

Divergent New Ideas	Phase I Composite Designs	Phase II Consolidated Designs	Phase III Stretched Designs

ROBUST DESIGNS

DESIGN FAMILIES

The initial design brief will often set terms of reference for the composite, consolidated and stretched design phase; as a consequence, a robust design brief should lead to a robust design while a lean brief will lead only to a lean design unless it is subsequently modified officially or unofficially, as sometimes happens.

Composite designs essentially filter out some of the potentially fruitful new ideas and lines of development (this process does not necessarily just bring together good ideas - sometimes there is a mix of good and bad).

Consolidated designs rationalize the filtered new lines of development and then emphasize and de-emphasize various ones into some sort of workable compromise in terms of makeability and hopefully profitability.

Stretched designs rework the workable compromise to arrive at new variants which are re-rated, up-rated and de-rated to cater better for existing or new markets. If there are two or all three re/up/de-rated versions, then there is a DESIGN FAMILY.

Composite and consolidated designs are convergent design processes, while stretched designs are a divergent process.

Composition, consolidation and stretching phases do not all have to be done by the same individual or group or even within the same organization. In the innovation literature there are lots of second-to-the-market firms who are good at the consolidation phase and/or particularly the stretching phase. Big aviation and automobile firms normally proceed through all three phases, but if competitors come up with something new which allows the whole process to be shortened, most firms will take it up and incorporate it into their own designs.

Robust and lean designs

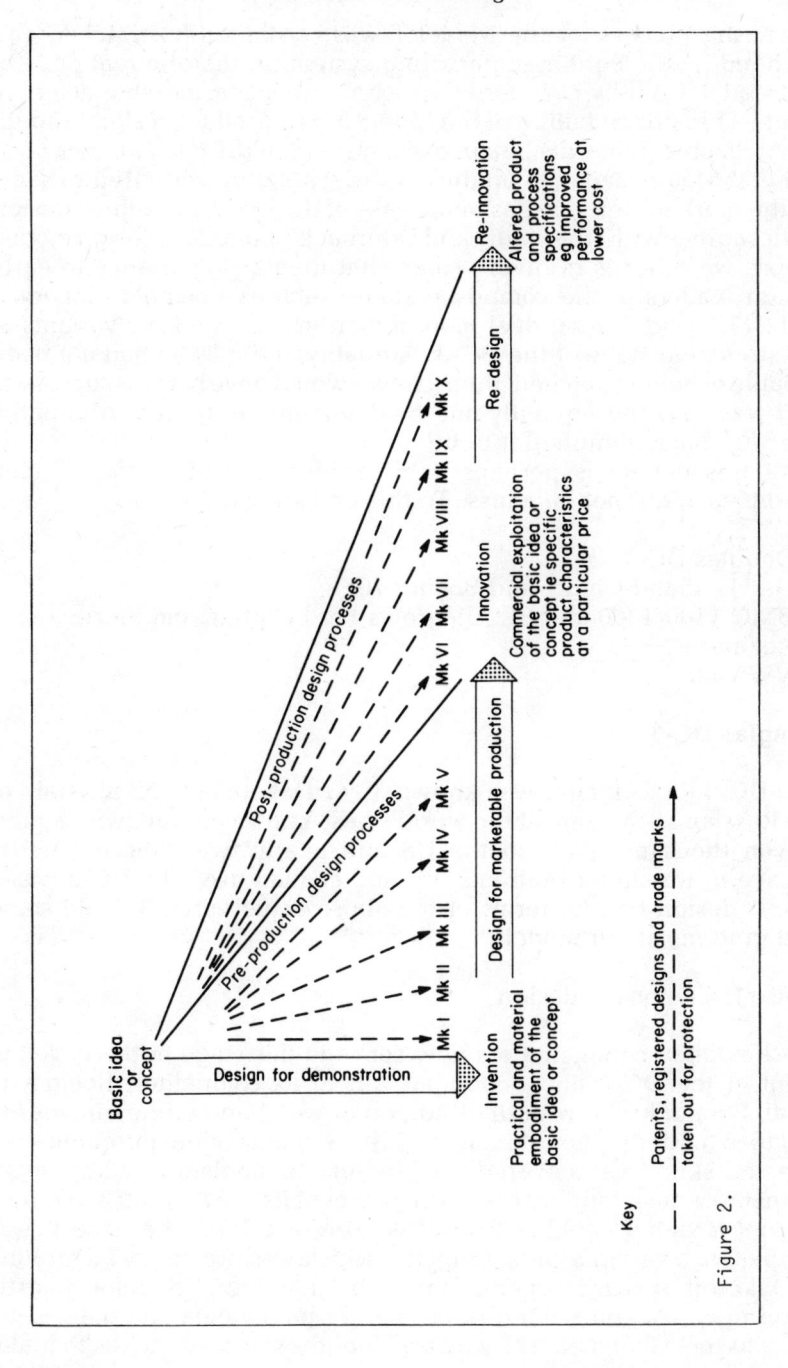

Figure 2.

ufacturing process for the Model T was considerably stretched and modified by the Ford manufacturing system, so that the real price of the Model T fell by two-thirds, which is still a considerable achievement.) This stretch failure of the Model T as a product design brought about its precipitous decline in market share in the late twenties once General Motors began to produce a variety of up/re/de-rated variants for the market. Certainly, as in the case of the DC-3, our robust design model agrees with Abernathy and Utterback's dominant design model. Where we differ is in giving somewhat greater importance to early design leaders at the composite stage, such as Boeing's Monomail and 247D, and a great deal more importance to the later variants at the stretching stage of the DC-3. Arguably, if the DC-3 had not been capable of being stretched, then it never would have been as successful as it was, and consequently not the dominant design that Abernathy and Utterback identified it to be.

All this debate is perhaps a bit academic, and a few detailed examples might not be amiss. To this end we will look at:

1. Douglas DC-3
2. De Havilland Comet and Boeing 707
3. BMC 1100/1300 and BL Allegro vs Ford Cortina and Sierra
4. Renault R5
5. VW Golf

Douglas DC-3

The DC-3 (or as it later was known, C-47 Dakota or C-53 Skytrain or the Russian Li-2) is one of the world's best-known aircraft, with almost eleven thousand made in the US and nearly two thousand under licence in Russia for both military and civilian use. The DC-3 was a robust design both in terms of the abuse actual aircraft could stand and in terms of our model.

Phase I: Composite design

Most aviation commentators now concede this stage in the development of the DC-3 not to Douglas but to its competitor, Boeing. In 1930, Boeing broke with the tradition of wood and fabric biplanes to produce and fly the Monomail 220, a monocoque duraluminium stressed-skin, cantilevered low-wing monoplane, with semi-retractable main landing gear. The power of its 575 hp Pratt & Whitney Hornet B radial could not be effectively employed because it was coupled to a variable-pitch propeller which was pre-set on the ground for take-off or cruising performance, but not both. The pilot was still in an open cockpit behind the space for passengers and mail. Later in 1930 the Monomail 221 was first lengthened by 8 inches to make room for two more passengers, and then by 2 feet 3 inches for another two passengers, to make room for a total of eight passengers.

Phase II: Consolidated designs

From the Monomails 220/221 and a derived scaled-up two-engined bomber prototype (Model 215), Boeing went on to produce the 247 and 247D, which became the starting point for all piston-engined commercial and cargo aircraft for the next 30+ years, be they two- or four-engined types, both for the military and the civilian markets. The 247s were produced to the requirements of Boeing's then parent company, United Aircraft/Airlines. This aircraft of 1933/4, powered by two 550 hp radials, could climb on one engine, and with both was 50 mph faster than other standard aircraft. There was wing and tail de-icing and a trim-tab system for ailerons and elevators. With a crew of two and a stewardess, ten passengers could be carried along with enough room for galley and toilet facilities. The only problem was for United Airlines' competitors, who could not get access to the tied production of Boeing 247s. Just over sixty 247/247Ds were produced before they were made obsolete by Douglas's DC series.

TWA led the competitors and called for outside tenders. Douglas's DC-1 proposal was accepted. John Northrop developed a multicellular structural system for the DC-1 which gave it and all its successors its well-renowned ruggedness. The DC-1 of 1934 with twin 700/710 hp supercharged radials had a better range than Boeing's 247, cruised 35 mph faster and carried two more passengers, for a total of twelve. Before production began, the DC-1 was enlarged to the DC-2, with 720 hp engines and fourteen passengers. Around two hundred DC-2s were produced before the DC-3 arrived.

Phase III: Stretched designs

At the request of American Airlines, the DC-1/DC-2 series was stretched by lengthening the wings and by lengthening and widening the fuselage to produce the Douglas Sleeper Transport, or DST. Beginning in 1935/6, the DST could fly transcontinentally across the US in stages, with berths for sixteen people. It was then realized that by replacing the berths with seats, the capacity, when compared with the DC-2, could be increased by 50% to some twenty-one passengers (not all re-ratings are planned for in a robust design). Range, speed and loads for the DC-3 were steadily increased both before and after the war by the introduction of a succession of supercharged and turbocharged radials, going from 900 to 1200 hp. Although some strengthening was required, the DC-3 was a very robust design which easily accepted this up-rating. During and after the war, these stretched DC-3s were mainly used for cargo transport. A few were

148 Design, innovation and long cycles

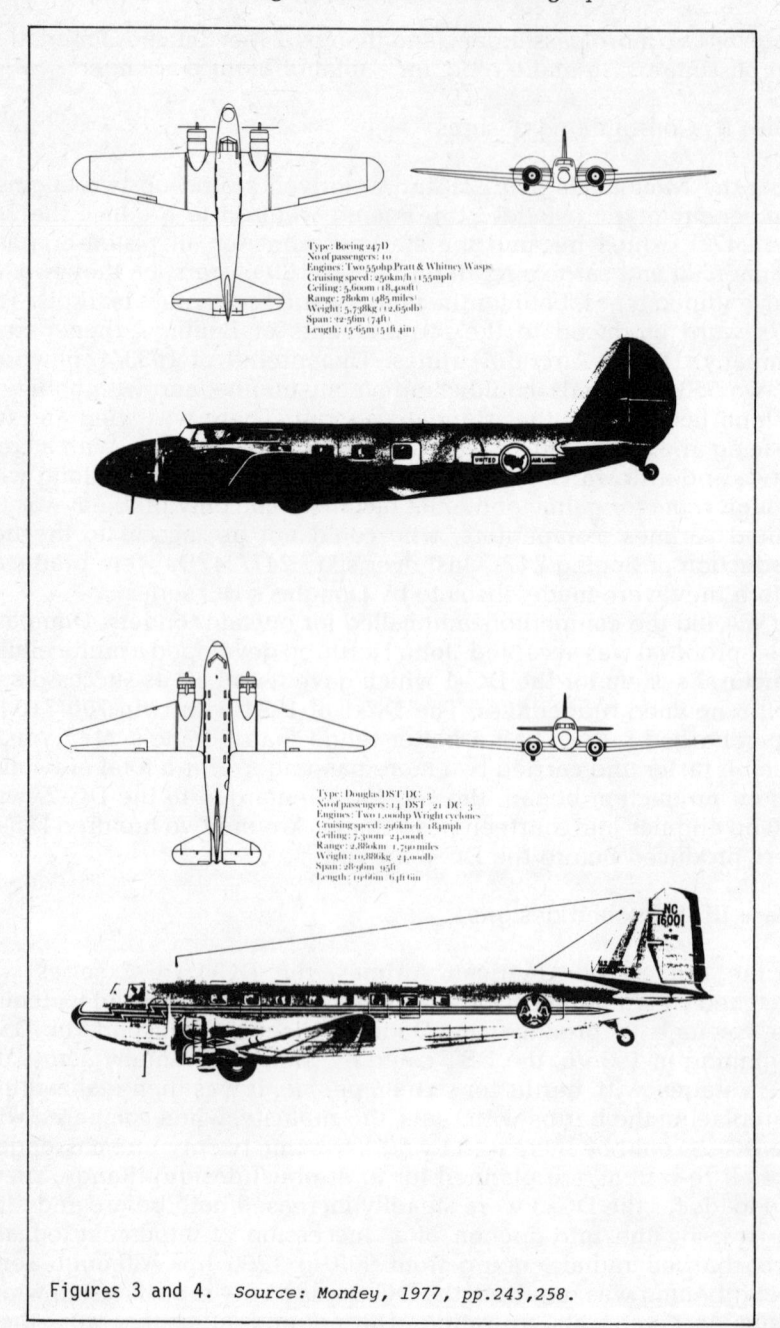

Figures 3 and 4. *Source: Mondey, 1977, pp.243,258.*

fitted with floats for seaplane use. As a casualty of the Pacific war, one DC-3 became a DC-2-1/2 because after one wing was damaged a DC-2 wing was substituted.

The biggest stretch of the DC-3 was its scaling up to become a four-engined DC-4 (or C-54 Skymaster), and then a succession of re-designs and re-engined versions which progressed with new wings and fuselages through the DC-6 series and the DC-7 series. The last of the DC-7 series in the late fifties was a long way from the DC-3 or DC-1/DC-2, but there is a path of nearly continuous evolutionary change linking them all together. (This design path was only possible because of the continuous parallel development of larger and more powerful radial engines.) Nevertheless, the amount of development or stretch was quite appreciable. For instance, a 1936 DC-3 with two 900 hp radials had a maximum take-off weight of 24,000 lb, cruising speed of 185 mph and a range of 1,500 miles. By comparison, and several decades later, a 1960s DC-7 (late series) with four 3,400 hp turbo compound radials had a maximum take-off weight of 143,000 lb, a cruising speed of 360 mph and a truly intercontinental range of 4,250 miles.

De Havilland Comet and Boeing 707

De Havilland with its Comets had several notable achievements. A Comet prototype first flew in 1949 and a production model went into service with BOAC to Johannesburg and Colombo in May and August of 1952 respectively. A production Comet beat a Boeing 707 into transatlantic service by three weeks in late 1958. In spite of the successes, and if one sets aside the problem of metal fatigue which interrupted production from 1954 to 1957, the Comets were an example of a lean design because of a lean design brief, while Boeing's 707 was an example of a robust design because of a robust design brief.

The pattern of development of post-war commercial aircraft was largely determined by the Brabazon Committee which was formed during World War II. The Committee, perhaps because turbojets were then an unknown (and so were the development costs), was not very optimistic, and its Type IV recommendations were for only a medium capacity, medium range turbojet aircraft. De Havilland wanted to produce a bigger, longer range turbojet for Empire routes which could compete commercially with more conventionally designed aircraft. BOAC and de Havilland eventually agreed a passenger jet design brief and two prototypes (under Ministry of Supply specification 22/46) were developed and completed in 1949/50. There followed a production Comet 1, with 4,450 lb thrust, Ghost 50 Mk1 turbojets and a range of 1,750 miles, which could seat 36 passengers. In 1952/3 the Comet 1As with Mk2 5,000 lb thrust turbojets could seat 44. Then

150 Design, innovation and long cycles

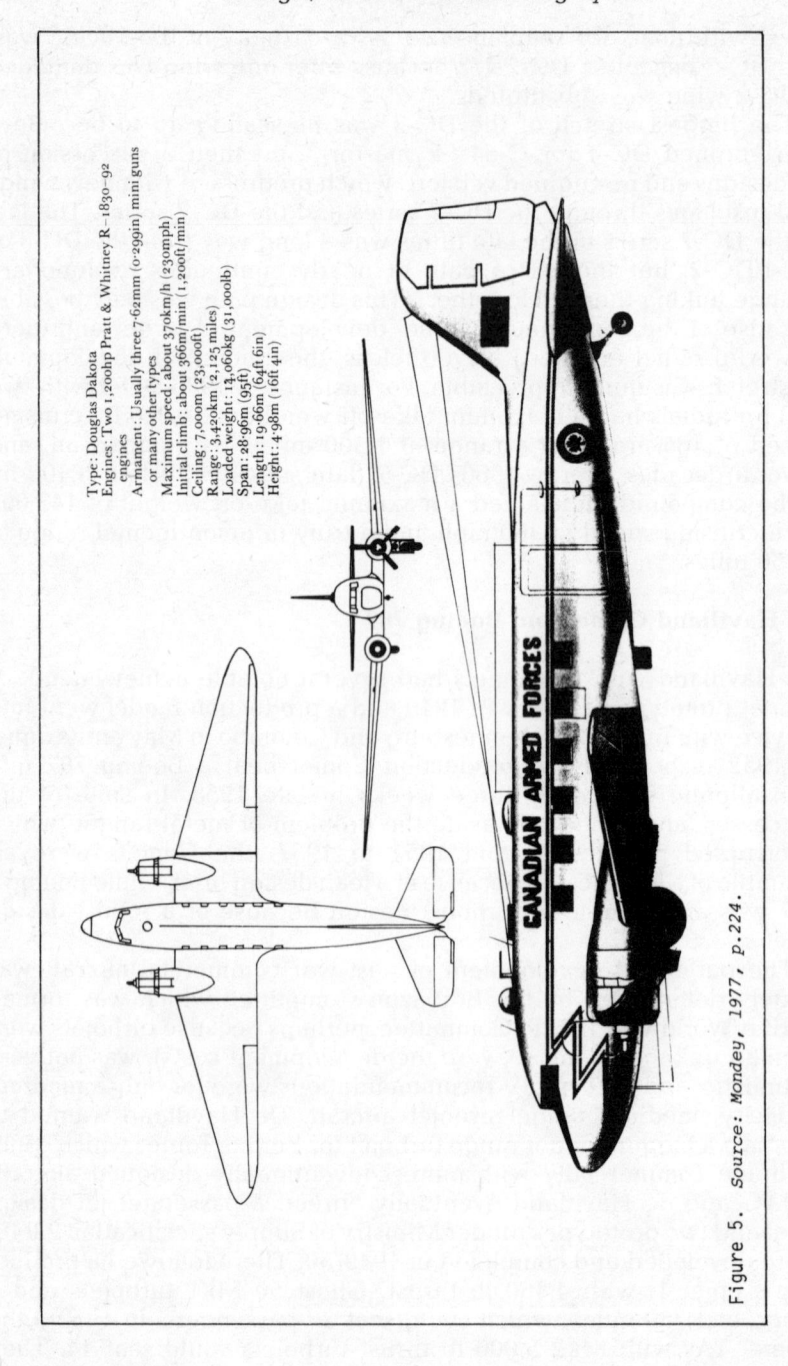

Type: Douglas Dakota
Engines: Two 1,200hp Pratt & Whitney R-1830-92 engines
Armament: Usually three 7·62mm (0·299in) mini guns or many other types
Maximum speed: about 370km/h (230mph)
Initial climb: about 366m/min (1,200ft/min)
Ceiling: 7,000m (23,000ft)
Range: 3,420km (2,125 miles)
Loaded weight: 14,060kg (31,000lb)
Span: 28·96m (95ft)
Length: 19·66m (64ft 6in)
Height: 4·98m (16ft 4in)

Figure 5. *Source: Mondey, 1977, p.224.*

Robust and lean designs

Type: de Havilland Comet 1
No of passengers: 36
Engines: Four 2,018kg/4,450lb thrust de Havilland Ghost turbojets
Cruising speed: 789km/h/490mph
Ceiling: 12,190m/40,000ft
Range: 2,816km/1,750 miles
Weight: 47,627kg/105,000lb
Span: 35.05m/115ft
Length: 28.35m/93ft

Type: Boeing 707
No of passengers: 179
Engines: Four 5,625kg/12,500lb thrust Pratt & Whitney JT3C turbojets
Cruising speed: 917km/h/570mph
Ceiling: 12,800m/42,000ft
Range: 4,949km/3,075 miles
Weight: 116,573kg/257,000lb
Span: 39.9m/130ft 10in
Length: 44.04m/144ft 6in

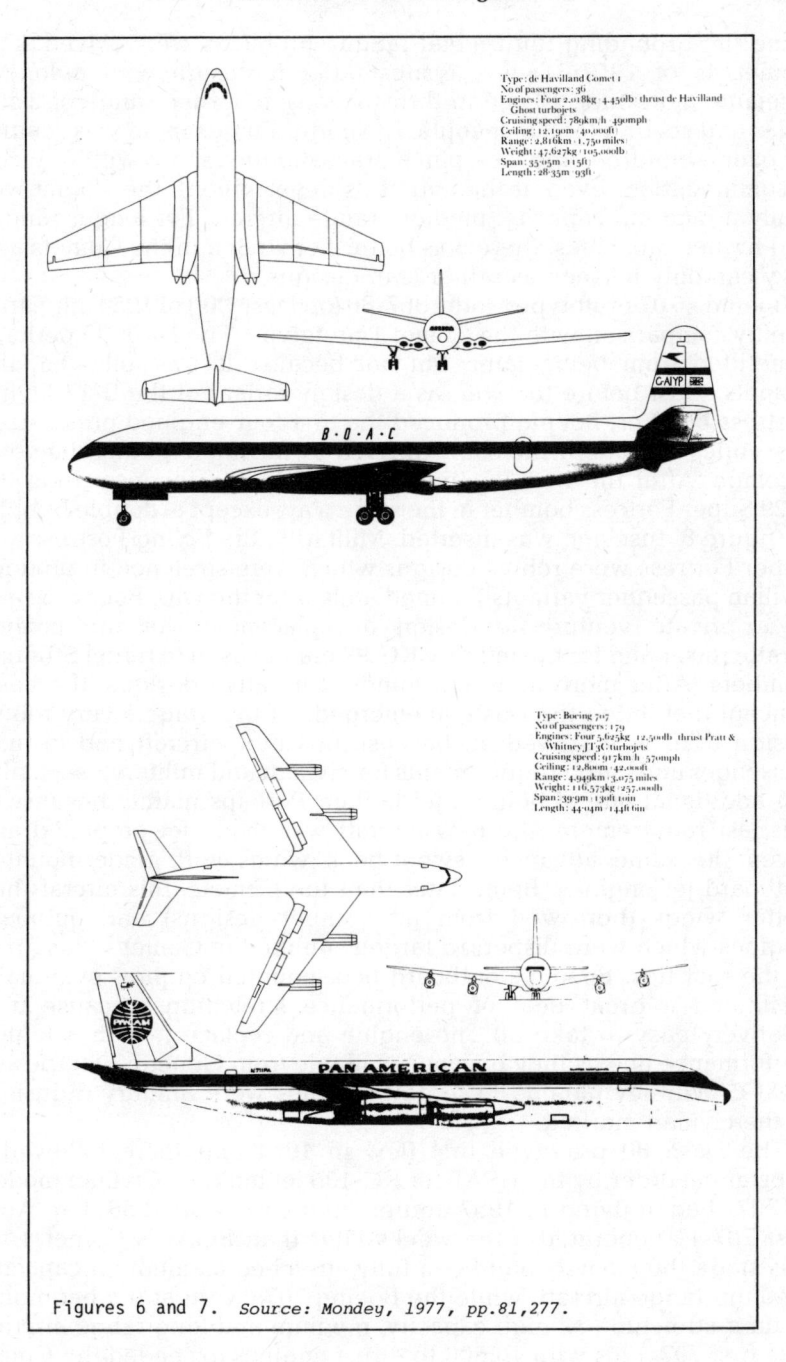

Figures 6 and 7. Source: Mondey, 1977, pp.81,277.

came the grounding until metal fatigue problems were solved. The Comet 4s of 1957/8 were a substantial re-design with a longer fuselage, more fuel, 10,500 lb thrust Avons, a normal range of 2,500 miles and seating for 80+ people. At shorter ranges there was seating for over a hundred, and, at a pinch, transatlantic ranges with reduced accommodation. Even at the end of its development, the Comet was really a medium capacity, medium range aircraft. For longer ranges and higher capacities, there was no further stretch in the Comets and they can only be seen as rather lean designs.

Boeing's 707 prototype Model 367-80 (or 'Dash 80') of 1954 appeared late by comparison with the Comet 1 prototype. The Dash 80 perhaps benefitted from being later, but not because it was following the Comets. (Just before the war, as a design variant of the B-17 Flying Fortress bomber, Boeing produced the first four-engined pressurized passenger airplane, which only differed by having a greater diameter fuselage. After the war, Boeing produced the Stratocruiser from the B-29 Super Fortress bomber in the same way, except a 'double bubble' or 'figure 8' fuselage was inserted. Militarily, the Flying Fortress and Super Fortress were robust designs which were stretched to produce civilian passenger variants.) Immediately after the war, Boeing began as a private venture to design a replacement for the civilian Stratocruiser and for the military KC-97 mainly used to refuel Stratojet bombers. After more than one hundred iterative designs, the basic concept that led to the Dash 80 emerged. At this stage a very robust design brief was arrived at, because this new aircraft had to fulfil passenger and cargo requirements for civilian and military users, plus the additional one of being a jet tanker. Perhaps mainly because of this last requirement, the new aircraft was made jet propelled and given the same advanced swept-back wings with undermounted outboard jet engines. Being later than the Comets, this aircraft had better wings (borrowed from jet bomber designs) and outboard engines which were dispersed targets, unlike the Comet's clustering at the root tips. Boeing's outboard pod-mounted engines eventually facilitated a great deal of performance stretching, because it is relatively easy to take off one engine and replace it with a higher performance one as they become available from General Electric and Pratt & Whitney (again, these new engines were military diffusions to the civilian market).

The Dash 80 prototype first flew in 1954 and there followed a substantial order by the USAF for KC-135 jet tankers. Civilian models or 707s began flying in 1957 and entered service in 1958. Pan Am's first 707-120 operated three weeks later than BOAC's Comet 4. At this stage the Comets had been fully stretched as medium capacity, medium range aircraft, while the Boeing 707s were at the beginning of their stretching as high capacity, medium and long range aircraft. The first 707-120s with 13,500 lb thrust engines exceeded the Comet

4s, which had only 10,500 lb thrust engines. Shortly thereafter 707-120s went to 17,000 or 18,000 lb thrust engines, and to 15,800 lb thrust engines and reduced weight in the 'high and hot' 707-220 variant. This increased power was used for increased capacity, mainly on medium range work. By January of 1959 the truly intercontinental 707s with longer wings began to arrive: the 707-320 and 707-420 (the latter with 17,500 lb Rolls-Royce Conway engines). About the same time, derived designs based on a shortened 707-120 began to appear as the intermediate range Boeing 720 series, with improved airfield capabilities. From 1960 onwards, new and some old 707s and 720s were equipped with higher thrust and performance American-made turbo fan engines. By the end of the seventies, more than one thousand 707s and 720s for passenger, cargo and quick change models had been produced.

Further variants out of the original Dash 80 robust design brief were to follow. At the beginning of the sixties, with three rear-mounted engines, there was the Boeing 727 series for short/medium ranges, and in the mid-sixties, with only two undermounted wing engines, there was the Boeing 737 series for short haul work. Both had new high light wing devices which gave good airfield capabilities even at second-class airports. Total production for Boeing 737s and 727s now exceeds two thousand. More interestingly from a design point of view, there is more than 60% commonality between the 727s and 737s, and both have the same fuselage cross-section from the floor upwards as the original 707. (Boeing's current 757, as the replacement for the 727, has more than 40% shared commonality with the forthcoming 767; the 757 and the 767 are the result of a joint design development program.)

While the 727 and 737 series are design variants, they still took a great deal of design and re-design work. Up to three hundred design studies for the two were undertaken, with perhaps one quarter going through to the wind-tunnel stage. The stretching phase is no more easy than the composition or consolidation phases.

The previous story was repeated at the end of the sixties, when the 707 design concept was more than trebled in capacity to produce the 747 design. Boeing was convinced that passenger and cargo versions would be equally important. Again, more than one hundred design studies were undertaken in the scaling up exercise. No prototype was needed, and the first production model flew in 1969 and entered service at the beginning of 1970. Almost six hundred 747s have now been sold as passenger, cargo, quick change and special purpose variants. In part, these variants were designed in by Boeing at the consolidation phase in order to facilitate their derivation in the stretching phase. Both the original Dash 80 (or KC-135 and 707) and the 747 are robust designs with large design families allowing for substantial re-rating and the production of variants, e.g. 720s, 727s,

154   Design, innovation and long cycles

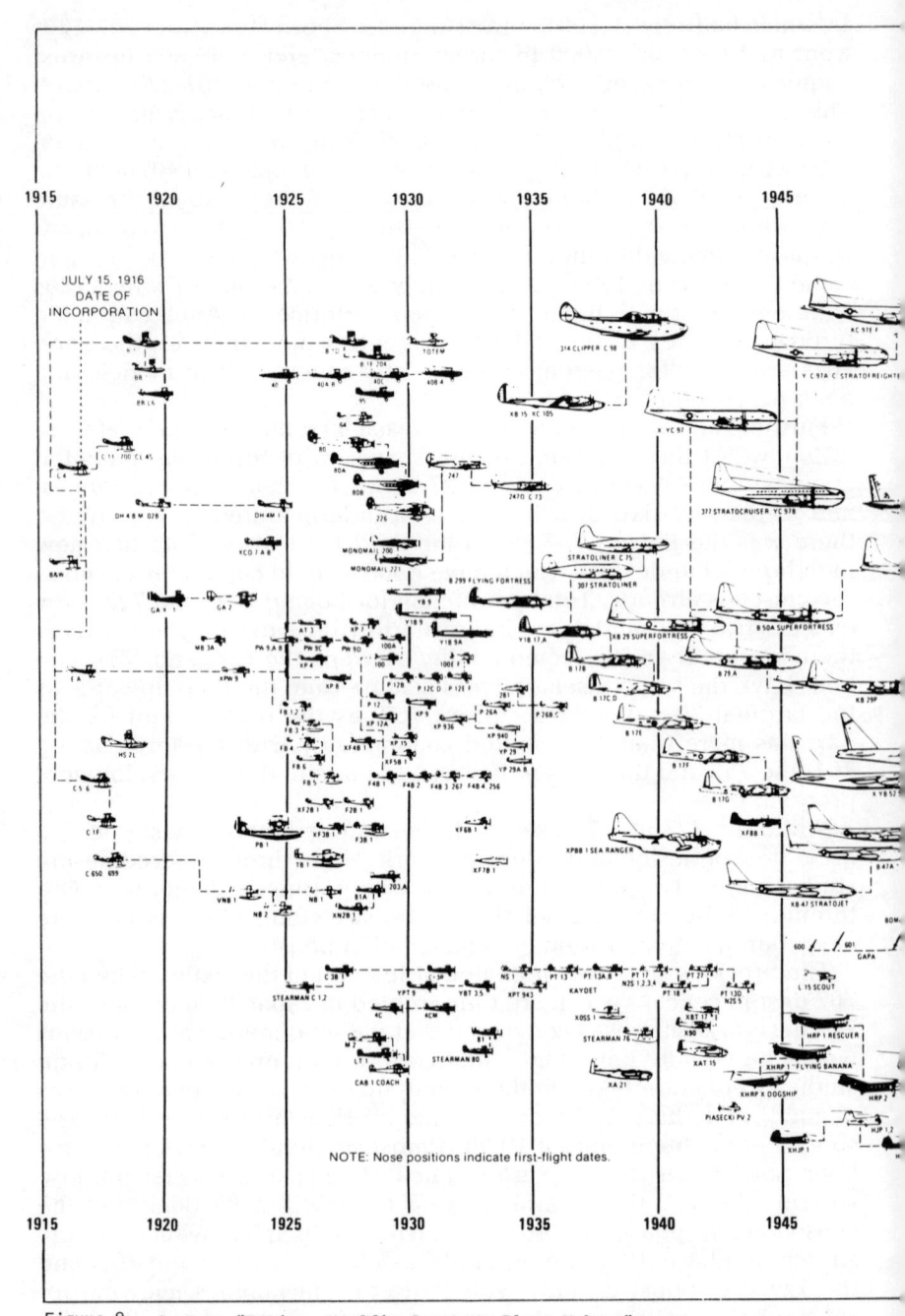

Figure 8. Source: "Boeing: World's Greatest Plane Makers",
S. Butterworth (ed), 1982.

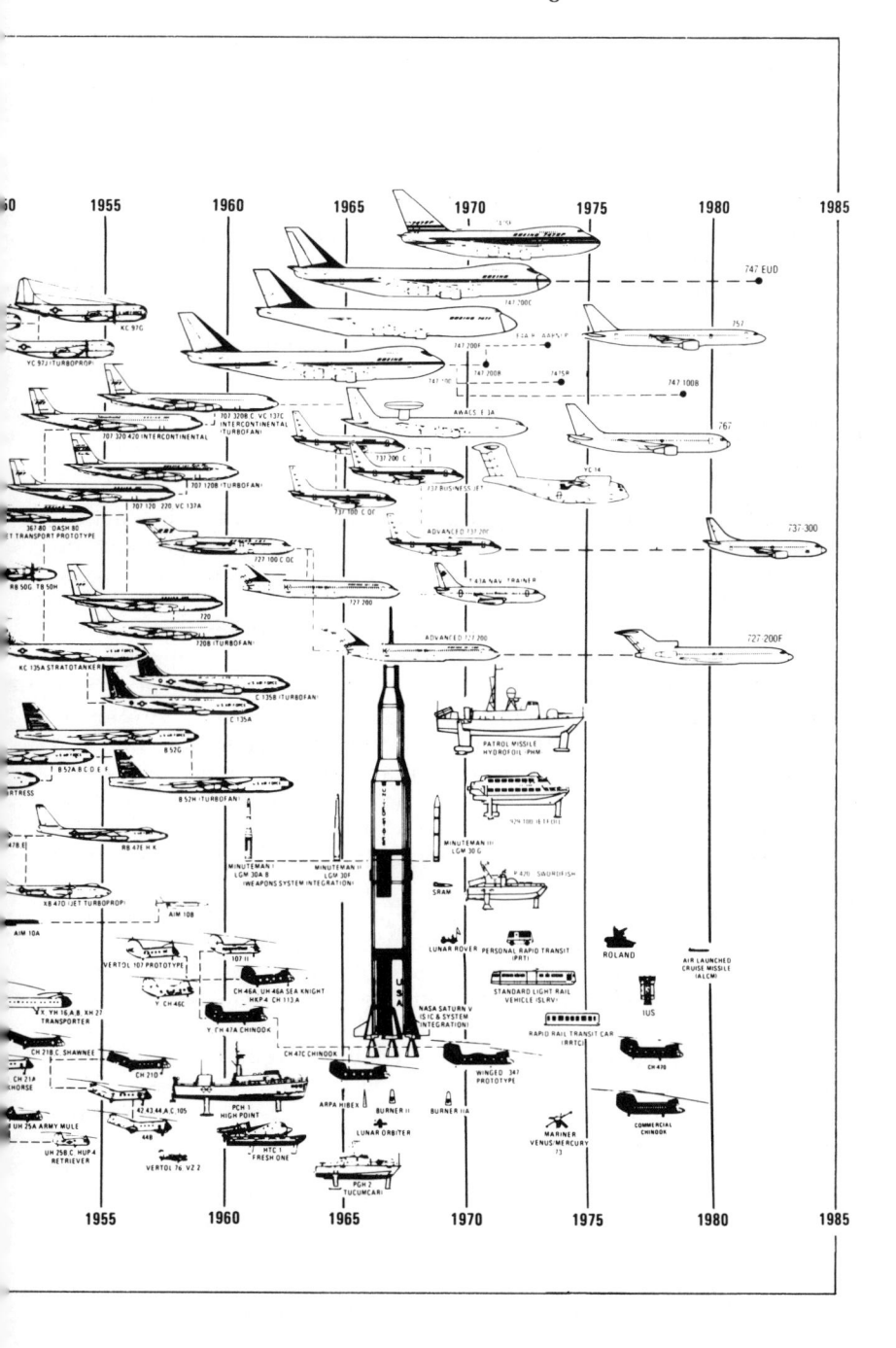

Robust and lean designs

737s and 747SRs or 747SPs. The biggest difference in the intervening twenty years between the Dash 80 and 747 design concepts was the introduction of CAD. CAD is now very much a part of aircraft design study exercises leading to more robust designs and to fewer or no prototypes being needed.

While Boeing's 707 and 747 are robust design examples and de Havilland's Comet is a lean design example, the British have not always had failures. This paper has not been looking at the short-haul market, where, in their own ways, the Vickers Viscount and BAC One-Eleven might be considered to be robust designs. In a paper by Rothwell and Gardiner (1983) in 'Design Studies', there is a section that looks at the Rolls-Royce RB-211 design family, which is another good example of a robust design.

BMC 1100/1300 and BL Allegro vs Ford Cortina and Sierra

With the introduction of BMC's Mini in 1959, there was the practical demonstration of the importance of the transverse engine/transmission radical innovation as an automotive propulsion system for a mass-produced car.

The BMC 1100 of 1962 actually had two radical innovations which eventually reshaped that class of car designs. Everyone always immediately notes the transverse engine/transmission radical design grafted over from the Mini, but most fail to notice the 'Hydrolastic' suspension which was another radical design change at that time and later refined to become what is known as the 'Hydrogas' suspension system used until recently in most BL cars. The Mini had a four-wheel independent rubber suspension which Moulton reworked on the more radical principle of gases and liquids under pressure for the at least 4 cwt heavier 1100. In terms of patents, the radical propulsion system developed quite differently from the radical suspension system. Issogoni's transverse engine and transmission design is basically captured in two patents - one for the layout of the engine and gearbox and one for the sub-frame for its mounting. A few years later there is one additional significant patent, where the manifolds are modified so that the engine/transmission unit could be rear mounted for a similar configuration to that found in some Fiat cars. Curiously, BMC does not seem even to have made a prototype with the rear-mounted transverse engine/transmission design.

Moulton, while working closely with BMC, was essentially an independent designer and engineer. His radical design for the Hydrolastic/Hydrogas suspension has almost a dozen patents covering it. The radical transverse engine/transmission design seems to have been conceived of and working out in one major step, while the radical hydrolastic suspension was designed and re-designed in a series of small steps. There are obvious differences between Issogoni and Moulton and the resources each had at his command, but the main lesson is that equally radical technical changes can be achieved with a large or small number of steps in the design process.

The year 1962 not only saw the launch of the new radically designed BMC 1100, but also the new conventionally designed Ford Cortina Mk I. In addition, the Consumer's Association took the radical step of publishing 'Motoring Which?' as the first of what was to become a whole series of specialist 'Which?' reports. Unlike the motoring trade press, 'Motoring Which?' does not carry advertising and tries to maintain an independent critical view of the product on offer by motor manufacturers. Fairly early on in the history of 'Motoring Which?', in 1963, the new 1100 and Cortina were judged to be 'joint best buys' over the also new Renault R8 and the older Hillman Minx and Wolseley 1500, in terms of performance, handling and economy. Thus in 1962 we start off with one conventional and one radically designed car for the medium class from each of Britain's two major car firms, and yet twenty years later only one can be said to have been a success - and it was the more conventionally designed car. The last Cortina of 1982, while definitely showing its age, was consistently better in 'Motoring Which?' reports than BMC's 1100/1300 and later Allegro and Marina. Over twenty years, the Cortinas were not always best when compared with foreign makes - but definitely better than those from BL. The question naturally arises as to why the conventional design turns out better over the long run than the radical design. The short answer is that the last Cortina of 1982 was not the same car as the Cortina Mk I of 1962 - except in name only. Ford continually evolves and re-works its car designs for each mark number, while BMC/BL 'locks up' on its designs. The mark number versions of the 1100/1300 and the Allegro often embody the most minimal design changes. The lesson seems to be that the new radical or new conventional designs can be equally important at the beginning of a new generation of cars, but for long-term success one has to keep working at improving the basic design of either type. The original Volkswagen 'Beetle' pre-dates the success of Ford's Cortina, but it also demonstrated the same lesson, except for starting with a radical rear-wheeled air-cooled propulsion system. Over more than 25 years, Beetle designs were worked and re-worked until, like the Cortina, the only thing that remains the same is the name. Even those Beetles currently being made in Brazil are still being re-worked.

As one simple but crude measure of design change, vehicle weights can be observed. The first Cortina started off at 15.5 cwt, while a Cortina 80 started off at 19.7 cwt. These were the minimum weights for the most basic model. Top line models could be up to 6 cwt heavier, depending on the options or extras which had been added. By way of contrast, the 1100/1300 and Allegro series both started off at 16.7

cwt and never added more than about 1.5 cwt for the heaviest model. Given that extras and option packages add weight and, more significantly for the manufacturers, profits - then Ford's Cortina series consistently offered a potentially more profitable model mix than that of BMC.

Since the 1100/1300 and Allegro series were essentially locked-up designs for each of their production runs, we will look at Ford's Cortina in somewhat more detail to see just what was being done as the Cortina began 'putting on weight' during the twenty-year production run.

See Table 1 for specifications of the Cortina Mk I, Mk II and 80, and for the Sierra. Obviously, not all details could be included; hence we have tried to give a representative selection only. Also, we have only given data for the first two basic and most popular models in each Cortina or Sierra series. The importance of other versions with different engines is taken up next.

Culshaw and Horobin in their 'British Cars' (1974, p.134) have concluded that Ford's greatest strength was in design depth and discipline, which allowed for change (but not too much at a time) and for it to remain in step with the buyers in the sixties and early seventies who wanted an increased range of models, prices and performance characteristics. It could be urged that this design depth and discipline continued into the late seventies and through to the present. The only difference is that energy efficiency and aerodynamics have been added to the buyers' range of product specifications. The Sierra has basically the same dimensions as the Cortina 80, but with two important exceptions: (i) new modified polycarbonate impact-resistant bumpers that increased the overall length while the basic bodies of the two are still nearly the same, and (ii) the Sierra is 0.2 cwt lighter than the earlier Cortina 80. While a 0.2 cwt weight loss is not much, it is an achievement because the Sierra specifications are an improvement on those of the Cortina 80 - it is a case of more for less, which is a definite improvement in design terms.

Perhaps one can be a bit more specific about the overall progress in the state of the art for a medium sized car with two examples; one looks at the vehicle over all, and the other at a particular sub-assembly:

1. The Cortina Mk I of twenty years ago had dimensions and performance figures more similar to those of a current Escort than to those of a Sierra, and even when compared to the Escort its performance specifications have been improved. A Cortina Mk I and a current Escort have about the same weight of 15.5 cwt. While the Cortina Mk I was longer, the Escort is wider, with only a slightly smaller boot. In addition, a new Escort has better accommodation, acceleration, handling and fuel efficiency. What has happened in the past twenty years is that the so-called medium sized car has moved

TABLE 1

Evolving 'State-of-the-Art' Product Specifications

Specification (for Saloon Model)	Cortina Mk.I 4 Passengers	Cortina Mk.III 4/5 Passengers	Cortina 80 4/5 Passengers	Sierra 4/5 Passengers
●Dimensions				
Wheelbase	98"	101.5"	101.5"	102.7"
Track	49.5"	56"	56.7"	57.2"
Length	168.5"	167.8"	170.9"	173"
Width	62"	67"	66.9"	67.7"
Boot	13.9ft^3	12ft^3	11.8ft^3	12.5ft^3
Weight	15.5 cwt	19.5 cwt	19.7 cwt	19.5 cwt
●Engine and Performance (1st Model)				
Capacity	1198 cc	1298 cc	1298 cc	1294 cc
Type	ohv 1cyl.	ohv 4cyl.	ohv 4cyl.	ohv 4cyl.
Compression Ratio	8.7:1	9.0:1	1.2:1	9.0:1
Power	48.5 hp at 4800 rpm	57 hp at 5500 rpm	60 hp at 6000 rpm	60 hp at 5700 rpm
mph (max)	76	85	82	94.4
0.50 mph	14.8 sec.	13.3 sec.	17 sec.	14 sec.
mpg	30	24.8	30	30.7
●Engine and Performance (2nd Model)				
Capacity	1498 cc	1599 cc	1593 cc	1593 cc
Type	ohv 4cyl.	ohv 4cyl.	ohv 4cyl.	ohv 4cyl.
Compression Ratio	8.3:1	9.0:1	9.2:1	9.2:1
Power	59.5 hp at 4600 rpm	68 hp at 5200 rpm	74 hp at 5300 rpm	75 hp at 5300 rpm
mph (max)	80.8	92	93	102.5 [E100.6]
0.50 mph	12.8 sec.	10 sec.	10.8 sec.	11.0 [E11.1] sec.
mpg	27.2	24.6	25	27.9 [E31.7]

Note: Special Economy Version in [E...]

up from the general design specification levels that are like those of the current light car class. Most obvious is that the Cortina 80 and the Sierra are five-passenger cars, while the Cortina Mk I and the Escort are four-passenger vehicles. This progressive and continuous improvement of design specification levels should not be overlooked. The design terms are always being up-rated and revised; nothing is very permanent about car classes. BL has had 'locked up' designs which failed to keep up with this continual change of design class specifications. This has made it difficult to keep up with competition as new mark numbers appear and new generations of cars arrive.

2. In design terms, vehicle suspensions behave best when the unsprung weight is at a minimum and when each wheel is independently sprung. Without going into a lot of detail, there has been a steady improvement from the Cortina Mk I to the Sierra. While the track width was increased about 7.5 inches, which should have increased the stability but at the same time added weight, this was offset by other design changes. The Mk I had a solid rear axle and underslung leaf springs. The Mk III still had a solid rear axle, but with trailing arms and coil springs. The Sierra now has semi-trailing independent suspension. This design had been pioneered by Ford a decade earlier in its Granada/Consul series in the large/executive class cars. Even more up-market at that time, it was being used by Mercedes Benz. This sequence is typical of many improvements which are first introduced in up-market cars and then moved down through the car classes so that each class is progressively up-graded. The introduction by Ford of a semi-trailing independent suspension for a medium class car was not just an off-the-shelf re-drawing of an old design. The pivot angle was changed to improve power on/off handling, while taking best advantage of current steel-braced radial ply tyres with fuel efficient characteristics. Furthermore, the old Granada-type design would have increased costs and added 0.25 cwt in weight. The weight and cost penalties were avoided wtih a new lightweight differential design and a new tubular subframe assembly for the whole of the rear suspension.

Both of these examples illustrate the sometimes difficult art of getting more but spending less. This achievement is only made with even greater design expertise, be it at the level of a sub-assembly for suspension or the whole vehicle configurations.

Generally speaking, Ford has produced robust designs while BL has had lean designs. For instance, the Mk I Cortina had three engines and four horsepower levels, while the Mk I 1100 had only one engine and could only compete with the Cortina at the first level. Throughout the Cortina series, and even with the current Sierra models, there have been up to seven horsepower levels, while the best the 1100/1300 series and the Allegro series could do was three and four horsepower levels respectively. Most of the time BL cars only covered the bottom half of the Ford medium sized car horsepower classes. This was not an accident. Ford quite purposefully produced robust and generous designs which allowed for changes in engines between mark numbers and for a variety of horsepower classes within each mark number. The present Sierra has six different petrol engines and one diesel to cover seven horsepower classes. The only way to allow for this practically was to opt for a conventional front-engined rear-wheel drive design and not a transverse front-wheeled design.

The Mk III Cortina was virtually a new design when compared to the smaller Mk II and Mk I Cortinas. The Mk III had the now famous 'Coke bottle' shape. The Mk II and Cortina 80 progressively refined this into a squarer and boxier shape that the Cortina was to end its days with. Again, this was not an accident. In the patents literature of the early seventies, there is good evidence of Ford producing robust designs for the main structural components expressly in such a way that they could be re-skinned in a number of different styles. In the case of BL, there is no sort of patents evidence in this direction. The 1100/1300s and Allegros of this period had almost no skin changes and must be seen as very lean designs. Equally, the Marina (which was redone as the Ital) was similarly a very lean design. As part of the design brief which produced the Ital, almost no sheet metal changes were allowed. Partly, BL could not afford a lot of new dies, but also the underlying design of the structure did not allow for very much alteration. Some of the structural design considerations were still tied to those of the preceding earlier designs for the Morris Minor (itself a lean design, although in its time a venerable car).

Most people in Britain have tended to view BMC/BL as a technological leader, with innovations such as transverse engine/transmission, front-wheel drive and hydrolastic suspension. Ford's Fiesta of the late seventies was more than a decade after the appearance of the first Mini. While all these facts are correct, there are some other parts of the story which are not so well known. In the early sixties, Ford could have produced a car like BMC's 1100 with both a radical suspension and a radical propulsion system. While the Cortina Mk I was launched in the UK in 1962, in Germany in the same year Ford Werke launched the Taunus 12M. This Taunus 12M was a front-wheel drive car in the same class as the Cortina. The one big difference was that, unlike the 1100 or Mini, the engine was longitudinal. This design is not all that strange, because in 1982 the Japanese Tercel had exactly this arrangement. On the suspension side, the patents literature clearly shows that Ford worked out a pneumatic suspension which, at least on paper, was comparable to Citroen's or BMC's radical suspension designs. Given all this technical expertise and capability, why didn't Ford produce a car like BMC's 1100 with two major radical innovations?

The answer seems to be that although you can design things and even build prototypes, you should not always continue with that line of development, for cost and performance reasons.

At the end of the sixties the various national operations by Ford were pulled together to form Ford of Europe. The Escort of 1968 was Ford's first 'European car', with co-ordinated production mainly in Britain and on the Continent. A year later this was followed by the first Capri and a year after that by the Cortina Mk III. The Cortina Mk III of 1970 replaced the Cortina Mk II and the Taunus 12M. The Mk III was practically a new car, although it retained the Cortina name in Britain and the Taunus name on the Continent. The Mk III with its 'Coke bottle' shape was a success, but curiously it does not owe all that much to the smaller conventionally designed Cortina MK II or to the radically designed Taunus 12M. With the Mk III, Ford could have produced a front-wheel drive car with a pneumatic suspension - two radical innovations - but did not. The pneumatic suspension wasn't reliable enough in cold weather or cheap enough to build. The longitudinal front engine/ transmission of the 12M seems to have been a costly stretch for Ford Werke to produce, and possibly beyond the less quality-conscious and more disruptive British Ford workers of that time. As a result, the Cortina/Taunus Mk III was a larger, five-passenger car with a conventional design but at a higher specification level. The specification level for the Cortina/Taunus throughout the seventies was set by what German buyers wanted rather than British buyers, because the former group was more demanding as far as Ford's marketing people could tell. As a consequence, the British Cortinas were probably always slightly 'over spec' and ahead of demand in the British market - not a bad thing during the seventies, as the market specifications for all car classes were being progressively revised upwards.

Renault R5

The R5 or Le Car is an example of a robust design that almost happened by accident. Prior to the R5, Renault had a reputation for rather staid, reliable and practical cars such as the R4. Fortunately, the stylist Michel Boué of Renault was given an interesting design brief from which was to spring a design that was almost unaltered by the time it went into production in 1972. Since then more than four million R5s have been made. Boué's design brief was for a stopgap model with a limited production and a lifetime of only five years. To this end all the 'mechanicals' were to be borrowed from existing R4 and R6 models, thus simplifying production, servicing and reliability. Boué had to design a car body for markets which Renault had not previously catered for. It had to be a compact, two-door, hatchback coupe that had style. Specifically the style had to appeal to three groups in the market: women, families needing a second car, and young people in general. With a long suspension travel and wheelbase, comfort was good for a car in this class. In addition, amongst the borrowed mechanicals there were fairly large slow-revving engines and high-ratio final drives which produced high fuel mileage results. Boué was able to package all of these elements into a car with the now famous R5 style which, like some other French designs, has almost become 'traditional' and so far has not needed any facelifts. A few early problems with heavy steering and engine noise were basically cured early on, but still draw some criticism. The biggest changes have only increased its market appeal. There is now a five-door version for larger families. The more recent Renault Gordini Turbos with almost twice as much horsepower and torque have 'boosted' the R5 into the sporty-performance market. (There are also super R5 Turbos of the factory rally team with a mid-mounted turbocharged power unit, but these are not generally for sale.) As a stopgap model with an interesting design brief, the R5 has turned out to be a robust design that seems to satisfy at least four groups in the market, along with having that elusive quality 'style'. Significantly, the R5's style has had a lot of fashion to it, and at the same time it has been functional and technically practical - a very interesting combination.

VW Golf

With cars, there is a real question about design convergence or divergence. If, say, aerodynamics becomes pre-eminent in the design brief, will all cars have similar banana-type shapes and only differ by their identification badges? The underlying problem is: What do consumers want and what can manufacturers produce at commercially attractive prices? In reality, there are a variety of consumer interests to be served, and the problem has a variety of design solutions.

In recent times, many major manufacturers have opted for producing a composite/consolidated robust design which leads to a family of designs to satisfy the requirements of at least three or more specific customer groups. VW's Golf is a very good example of this approach. The Golf design family (see Table 2) starts with a cheap basic car and then goes on to more expensive versions which offer either more luxury or sporty performance specifications (sometimes both). Somewhere in the mix there are now special 'E' or Economy versions with particularly good fuel efficiency. The VW Golf design family has the following elements and ranges:

* Doors - 3 or 5 or 2 with convertible versions
* Trim - 4 levels
* Engines - 4 petrol, 1 diesel; 1093 to 1780 cc

* Miles per gallon - 26.9/42.8 to 41.5/53.3
* Acceleration - 0-60 in 9.0 to 16.8 seconds
* Maximum speed - 84 to 114 mph
* Price - 4,156 to 7,990

In addition to these variations, it must be recalled that some engines, transmissions and floor pans are also shared with the Polo, Jetta, Scirocco and Passat model series. While Volkswagen has eight price versions of its Golf series, a home producer would refine this further. For instance Ford (UK), with its comparable Escort, has thirty-one price versions, more than covering the Golf design combinations and ranges. The Golf and Escort are both in the Light car class, and in one sense we have 'design convergence' in that Ford and Volkswagen produce only one car each in that class; but at the same time we have 'design divergence' because there are between eight and thirty price versions, trying to satisfy the range of interests of at least three main groups of consumers. Actually the situation is even more complex, in that the 5,000 to 8,000 price band permits consumers to switch from the light car class to the small, medium and large/executive car classes.

The Golf's powertrain units are also examples in their own right of robust designs. The diesel was the result of the conversion of an existing petrol/gasoline engine. Such a conversion was very attractive inasmuch as it allowed most of the existing tooling on the main engine line to be retained without substantial modification. If the original petrol/gasoline engine had been a lean design, it could not have been converted, as diesel engines are subject to much higher stresses. Having come up with a fuel-efficient diesel, it might at first glance seem a bit curious to produce a Formel E variant (using an existing petrol/gasoline engine but with a new transmission) that was also fuel efficient and 16% cheaper than the diesel version. The reason was that the diesel version only makes sense in terms of long-term operating economics in those countries where diesel fuel is at a much lower price than petrol/gasoline (as is the case in several continental countries). In the UK, where the prices are nearly equal, the Formel E variant is more attractive in terms of overall long-term operating economics. The Formel E transmission and final drive unit has an interesting combination of ratios that generally lowers engine speeds without sacrificing overall performance.

Overall, the Golf is an example of a robust design both in terms of body design and powertrain design.

A closing reflection

If one were to try to sum things up, then one might say that clearly robust designs are embodied in many good state-of-the-art examples

Table 2

Version	Engine cc/bhp	mpg urban/56mph	0-60 sec.	mph max.	Weight cwt.	Price
a) BASIC						
Golf C 3-dr	1093/50	30.4/40.4	15.6	87	14.7	£4156
b) INTERMEDIATE						
Golf CL 5-dr	1272/60	26.9/42.8	14.6	93	15.2	£4848
c) ECONOMY						
Golf C Formel E 5-dr	1093/50	40.9/54.3	13.2	84	14.7	£5574
Golf C diesel 5-dr	1588/54	41.5/53.3	16.8	88	14.7	£5345
d) LUXURY						
Golf GL 5-dr	1457/70	31.4/42.8	12.7	97	15.2	£5539 (Auto +£334)
Golf GL convt. 3-dr	1457/70	31.4/38.1	14.3	93	15.9	£7084 (Auto +£320)
e) PERFORMANCE						
Golf GTi 3-dr	1780/112	26.6/47.9	8.2	114	15.9	£6499
Golf GLi convt. 2-dr	1780/112	26.6/46.3	9.0	108	15.9	£7990

Source: "Autocar", October 1980.

in both the aviation and automobile sectors. While robust designs can occasionally happen almost by accident, most are the outcome of an earlier robust design brief. This earlier brief establishes a framework for combining new lines of technical development, and then consolidating them in the light of the prevailing economic priorities of prospective consumers at that time and place. Having achieved this consolidation, there are inherent elements in a robust design that allow it to be stretched in various up/re/de-rated versions as the technological and economic climate continues to evolve. In an unchanging world, designs would need to optimize only to the appropriate mix of technological and economic elements. But quite obviously our world is always changing, and the stretching phase of robust designs allows for their continual re-adaptation. With stretching, many cars and aircraft that have robust designs continue to be produced, often for up to seven years and sometimes beyond, before the need for a new robust design brief clearly emerges. The lifetime of a robust design need not be fixed. For instance, in the instrument industry, a robust design might have been good for four to six years, but more recently the provision of some micro-electronic computing capabilities has led to 'smart instruments'. Even if they are based on robust designs, these new smart instruments will probably only survive for two to four years. Equally, the lifetime of automotive and aircraft robust designs could be shortened if micro-electronics makes a similar substantial impact and if this new technology continues to undergo its own rapid rate of development.

For some, the 'over-engineering' that is characteristic of robust designs seems often to be something of a luxury. But this is only true if one looks solely at how appropriate a design is for the 'here and now'. A more realistic assessment accepts that the here and now is always changing, and it is the over-engineered robust design that allows for adaptation in an ever-changing world.

APPENDIX: MARCONI'S WIRELESS ROBUST DESIGN

Phase I: Composite Design

Marconi's first and perhaps biggest step was to bring together two major lines of development. Firstly, there was the 'hard wired' telegraphy communications systems which were essentially based on 19th century electromechanical technology. Secondly, there was the science-based electromagnetic theorizing of Maxwell and the empirical demonstrations of Hertz. Curiously, telegraphy people were not interested in wireless techniques, and scientific people were not interested in communications systems (scientists were primarily concerned with showing the similarities of electromagnetic radiation and light). Marconi clearly saw the need for a wireless electromagnetic communications system. Indeed, at the beginning he could not believe that someone had not already had the same idea, worked it all out, and found it to be a hopeless project.

Phase II: Consolidated Design

Very early on, Marconi made four major design advances which could all be linked together as a system:
(i) Separate receivers and keyed transmitters, each with their own antennae and grounding systems (this seems obvious to us now, but Marconi had to invent and design these essential elements);
(ii) Coherer detector sensitivity was improved, thereby allowing greater ranges;
(iii) Coherers were continually re-sensitized by electromagnetic tappers to permit coded sequences to be detected (the coherers used by scientists were essentially one-shot devices.

Phase III: Stretched Design (first few years only; much more followed)

During the first world war, hard vacuum valves/tubes became available and wireless communication began its electronics age. But before then, great advances had been made with electromechanical techniques. Here are but a few:

1. Around obstacles (e.g. hills) and greater than line-of-sight communications.
2. High-powered electrical motor-driven spark gap transmitters.
3. Large directional-antenna systems for receiving and transmitting.
4. Portable wireless systems for land, sea and air.
5. Ship-to-ship and ship-to-shore relay networks for news services, military use and emergencies.
6. New, more sensitive magnetic wire clockwork detectors, or 'maggies'.
7. Receiver and transmitter turning circuits.

References

W.J. Abernathy (1978) 'The Productivity Dilemma: Roadblock to Innovation in the Automotive Industry', Baltimore, Johns Hopkins University Press.
D. Culshaw and P.Horobin (1974) 'The Complete Catalogue of British Cars', London, Macmillan.
D.T. Jones (1983) 'Technology and the UK Automobile Industry', London, Lloyds Bank Review.
D. Downs (1982) 'The Passenger Car Power Plant: Future Perspectives', Ricardo Consulting Engineers, Shoreham-by-Sea.
D. Mondey (ed.) (1977) 'The International Encyclopedia of Aviation', London, Octopus Books.
K. Munson (1982) 'Airliners from 1919 to the Present Day', London, Peerage Books.
J.M. Utterback and W.J. Abernathy (1975) 'A Dynamic Model of Process and Product Innovation', Omega, vol.3, no.5.
K.G. Wilkinson (1976) 'The Technology and Economics of Air Transport in its Next Phase', Aeronautical Journal.

9 Long-range strategic planning in Japanese R&D

Ichizo Yamauchi

During the past ten years or so, which I believe to have been a critical and significant period for Japanese industries, I have had the opportunity not only to watch and study the changes in corporate behaviour and industrial structure in Japan, but also to interview the top management in various companies to get first-hand knowledge of their ways of thinking about the management of their companies.

During this research and study, I discovered several rather interesting facts. For instance, during the past decade, it seems that the concept of corporate and R&D strategy in Japanese industry has been changing quite rapidly. Roughly speaking, before the 1970s, and especially before the 1973 oil crisis, the big private Japanese companies failed to grasp the concept of corporate strategy and strategic R & D planning.

To take an interesting example of the way of thinking in private corporate management, when I met the president of a big enterprise in the electronics and electrical sector, he explained why his company had diversified into the unprofitable computer and nuclear sectors. He said that a large-scale company such as his should serve the national goal (that goal being to catch up with advanced countries) and therefore, as long as that technology and those products were essential to the Japanese industrial development process in the future, he felt reluctant to leave those business areas, however risky and unprofitable they might be.

If we may define 'corporate strategy' as a series of policies and means designed to achieve a business purpose, which is chosen on the basis of reasonable criteria of benefit and loss, such as return on investment, then we can say that Japanese companies had not adopted a strategic way of thinking. I think that throughout the long history of Japanese industrialization, industrial policy in the government sector and corporate strategy in the private sector have been influenced and distorted deeply by the existence of 'the national goal', apparently needed to catch up with advanced countries. In other words, of the advanced countries Japan has had a working 'textbook' for industrialization. This textbook has provided not only a national goal but has also carried many implications for the means to attaining this national goal.

However, since the 1970s and the attainment of Japan's national goal, industrial policy in the government sector and corporate strategy in the private sector have been changing. But the way for Japan to adjust to these new circumstances is one of the most controversial matters in Japan.

Needless to say, long-range strategic planning in Japanese R&D did not escape the influence of the working textbook. Owing to the existence of the working textbook, R&D activity in Japan has shown peculiar characteristics throughout the long period of industrialization. I discuss this in more detail below.

In my opinion, the differences in timing of stages of industrialization in countries' histories are a crucial factor governing the differences in the characteristics of industrial policy and structure and corporate strategy. When we discuss the differences in industrial policy and structure between Japan and other countries, we are sometimes inclined to emphasize the differences in traditional culture. However, among industrialized countries, the cultural difference is not as much of a vital factor influencing the form of industrial policy and structure and corporate strategy, as the distinction between leader countries and follower countries.

Industrial policy and R & D strategy before World War II

To discuss long-range strategic planning in Japanese R & D, I start by investigating how this strategic planning has evolved from its early stages. By taking a historical view, we are able to understand more easily the peculiar characteristics of the Japanese industrial process and R & D strategy that developed as a result of Japan being a 'follower' country.

In 1853, about 100 years after the industrial revolution had started in Britain, visiting American steamships caused quite a panic in Japan. The advent of such industrial realities forced Japan to abolish its national policy of isolation and to permit private individuals to begin trading with foreign countries. This was the first opportunity for the Japanese people to encounter face to face the actual results of the industrial revolution.

Seven years after this, a skilful craftsman succeeded in building a handmade steamship, despite the fact that he had no scientific or engineering knowledge and had to use traditional tools. He simply attempted to copy a foreign steamship in detail. His ship did actually work, but because of a lack of precision, practical use of the ship was out of the question. The government, which had ordered him to build the ship, realized that there was an unimaginable gulf between Japanese traditional artistry and modern Western technology, and that by now Western technology was heavily dependent on machines and not on the craftsman's skill. The intellectuals in Japan realized the importance of modern technology and felt that unless they could adapt to it Japan could not survive as an independent country. Therefore, from that beginning of industrialization up to the second world war, the first priority of Japanese political and industrial strategy was building up sufficient military force and using moderrn Western technology to compete with advanced countries.

I think that the handmade steamship did give one interesting indication - fortunately for Japan, the level of up-to-date Western industrial technology, as exemplified by the steam-powered locomotive engine based on iron materials, was not so far ahead that traditional Japanese knowledge could not follow.

It is worth remarking that the industrial technology which Japan faced in the 1850s, after a century of industrial revolution from 1750 to 1850 in Great Britain, such as cotton, coal, iron and steam-powered engine technology, had been formed on the basis of experiential knowledge and not on established scientific and technological knowledge, and that the great majority of industrial workers in Britain were skilled hand craftsmen or labourers, working in small workplaces or at home, while only a small part of the total labour force was working in mechanized factories.

By 1868, only 15 years after the first visit of the American steamships, the feudalistic Japanese government had failed in the attempt to adapt to Western technology and was overthrown in a political revolution. After a hard period of trial and error in the early stages, the new government managed to form a series of economic and industrial policies. The outline of the general policy can be summed up as follows.

Late 19th-century economic and industrial policies

First, the national goal of these policies was for Japan to survive as an independent country in the harsh international environment. In order to attain this goal, Japan's first priority was to secure enough military force as soon as possible and also, by means of effective industrial policy, to push rapid progress in their industrial revolution.

Second, along with this national goal, the important point was how such industrial policy was to be put into effect, viz how to supply sufficient money, labour and technology. Regarding the questions of money and the labour force, the agricultural policy presented the solution. By abolishing feudal ownership of land and forcing its privatization, the government tried to give incentives to increase agricultural productivity. On the other hand, the new government introduced a new tax system for farmers, as severe as the former feudal system, in order to raise a fund for industrialization and maintain the low standard of living.

Hence, partly because of this tax system and partly because of the substitution of the traditional farm-made products with sophisticated imported products such as cotton textiles and certain foodstuffs like sugar (in common with the phenomenon in the pre-industrial revolution period in Europe) lower-class Japanese farmers lost ownership of their land and were unable to support all the members of the traditionally large family. These families thus became a source of modern industrial labour and, additionally, the farmers' low standard of living set a standard for these workers' wage level. In this way, these lower-class farmers and their families played a significant role as a source of cheap labour. And since the new government established a good compulsory elementary education system in the 1870s, Japanese industry was able to secure a large enough cheap and modern labour force to compete with advanced countries.

Regarding the source of technology, the new government set up several state-owned companies in major industrial fields and depended on foreign sources for the whole range of modern technology for these companies, e.g. by importing machinery and production equipment and design of products as well as hiring a number of foreign engineers. After adopting this policy, the Japanese government and the intellectuals realized that this kind of technology transfer did not work well, partly because the cost of technology transfer was so great and partly because of a lack of sufficient knowledge and skill in Japan to adjust imported technology to Japanese industry needs. The imported technology of that period had been designed for use by British skilled workers, and it was therefore essential for Japan, which had not accumulated enough skilled workers by that time, to adjust imported technology for Japanese unskilled workers.

Importing technology

However, there is an interesting example where things proved fortunate for Japan. Japan was not in a position to develop alternative technology by herself, but found a solution to the tech-nology problem by importing next generation technology.

The Japanese cotton industry had started by introducing mules and looms from Great Britain. However, the spinning mules and associated cotton technology had been designed for the skilled labour and craftsmen of Great Britain and therefore, with the lack of skilled workers in Japan and of a capable management system for these workers in Japanese factories, the imported technology had not been working efficiently and companies could not realize a reasonable profit. In that period, in the 1880s, in order to run a factory based on mule technology, the proportion of skilled male workers in the total labour force had to be as high as 40%. Thus the problems of getting such skilled workers and managing them were crucial for the companies.

These problems were basically solved by introducing next generation spinning technology, viz the Ring spinning machine. For operation of Ring spinning technology, the proportion of skilled male workers among total employees could be reduced to less than 15%. By the 1890s, Japanese cotton companies were able to organize a manageable working system on the basis of unskilled female labour. And owing to this technological innovation combined with domestic improvement in technology, the Japanese cotton industry could grow into a dominant export industry.

New policy

In this way, the Japanese government and the intellectuals came to realize the importance of a kind of R & D activity, even in the transfer of technology, and that as long as they maintained the early and simple industrial policy of technology transfer, Japan would not be able to build up enough competitive strength to compete with advanced countries in both the domestic market and foreign markets.

About 20 years after the establishment of the new government, in the 1890s, the government was obliged to introduce a new industrial policy as a solution to this problem. It transferred the state-owned companies, with the exception of defence-related industries, to the private sector and gave them various types of support in technology and subsidies. With the advent of this privatization, each company could secure the freedom and incentive to rationalize its production systems and introduce homemade improvements in technology adapted to its somewhat unsophisticated market needs and to the low level of labour skills.

In addition to this privatization policy, the government tried to improve R & D capabilities in both the government and private sectors. For this purpose it organized several state-run universities and sent many students to Europe.

Japan's uniqueness

Why was it that among the follower countries in that period, such as China and India, only in Japan did such an industrial policy work effectively and was the industrial revolution able to make rapid advances?

In Japanese academic circles, attention has been focused on Japan's positive adaptability to modern technology in the early industrialized period as compared with other Asian countries which became involved in the international trade markets at the same time as Japan.

In seeking the main reasons for the excellent adaptability of Japanese society to technology, I would point to (a) the fully developed nationwide domestic commodities trade system, e.g. for rice, cotton, silk, tea, oil etc., which pre-dates the industrialized period, and (b) what I would term the matured feudal system. The matured feudal system meant that the mechanisms or rules of societal formation were dictated by ownership of land - a step ahead of society based on the kinship system. By contrast to Japan, society in other Asian countries was still not free from forms of the traditional kinship system, nor did they have a well organized nationwide commodity market. Because of the matured feudal system, Japanese society was able to go rapidly through a transitional period to capitalist-style society and flexibly adopted European capitalistic mechanisms and rules of societal formation.

By contrast, even in modern-day Taiwan for example, Japanese companies have had confrontations with Chinese business partners in their joint venture companies over basic management policy. Sometimes the Chinese partner is reluctant to expand business scale beyond his family's management capability. Handing over any management authority to people outside the family is not liked. For the successful introduction of modern production technology, it is crucial to organize a management system exceeding the limits of family membership and which corresponds to the technology. In the early stages of industrialization, the Japanese succeeded in achieving that.

The importance of a well-developed nationwide commodity market from the pre-industrialized era is self-evident. Because of the flexible market mechanism of this Japanese nationwide market, not only could imported commodities be readily diffused all over the country, but so too could imported modern technology and homemade technical innovations, and such technology would be the cause of a succeeding series of technical innovations.

For example, only a few years after the abolition of the national policy of isolation, on the one hand the traditional tea, silk and vegetable oil industries had become transformed into export industries and had rapidly increased their output, while on the other hand, without any need for political pressure, Great Britain was able successfully to penetrate and dominate the Japanese cotton textiles market, and the Japanese traditional cotton industry instantly suffered great damage. The general feeling is that the dynamism of the Japanese market which both these examples illustrate was directly dependent on the fully developed traditional distribution system and the manufacturing system organized by the big feudal merchants.

Exactly because of this market dynamism in Japan, both the government and industrialists could recognize the importance of modern technology and could find the key factors for success in each industry. For example, among the Asian countries, only in Japan was there a cotton industry which adjusted to imported technology and which developed remarkable homemade technical innovations in the industry. As a result of these adjustments at an early stage, companies were able to rid the Japanese export market of imported products and cotton grew into a main export industry in less than 20 years (see Figure 1).

During the early part of the industrialization process, certain peculiar characteristics of R & D strategy in Japan formed, and these characteristics were maintained until World War II.

R & D characteristics

First, by the second world war the national goal of R&D strategy was to build up a homemade production technology system by modifying imported technology. Generally speaking, almost all industries, especially in defence-related areas, including the steel, shipbuilding, aircraft and weapons industries, managed to attain this goal, albeit at tremendous cost.

Second, the homemade production technology system had to serve two different purposes in Japanese industry. The first was to allow production of first-class technological products, especially defence-related products, from raw materials to finished goods, without any dependence on imported technology. In the case of defence industries, which dominated the Japanese machinery industries at that time, the government and industrialists did not pay as much attention to reducing production costs or increasing productivity as to how they could modify imported technology to produce first-class defence equipment and adjust it to Japanese raw materials sources and to domestic production technology.

The second purpose, mainly in the commercial sector, was to facilitate a reduction in production costs by substituting cheap homemade production equipment for expensive imported equipment. The non-defense industries focused their R & D efforts to this end in order to strengthen their competitive power against overseas companies, in both the domestic and overseas markets.

For example, Mr Toyota, the father of the founder of Toyota automobiles, invented such an innovative automatic loom that he was able to sell the patent to British manufacturers prior to the first world war. But it was generally accepted that his great success in this venture rested on the remarkably cheap price of the product rather than on its superiority of performance or productivity in comparison with imported machines. Because of the cheap cost of labour at that time, the main goal of R&D strategy in the private sector was not to improve productivity but to reduce the cost of production equipment and to improve the ease of use for the Japanese unskilled labour force.

In this way, because of the separation of purpose of R&D activities

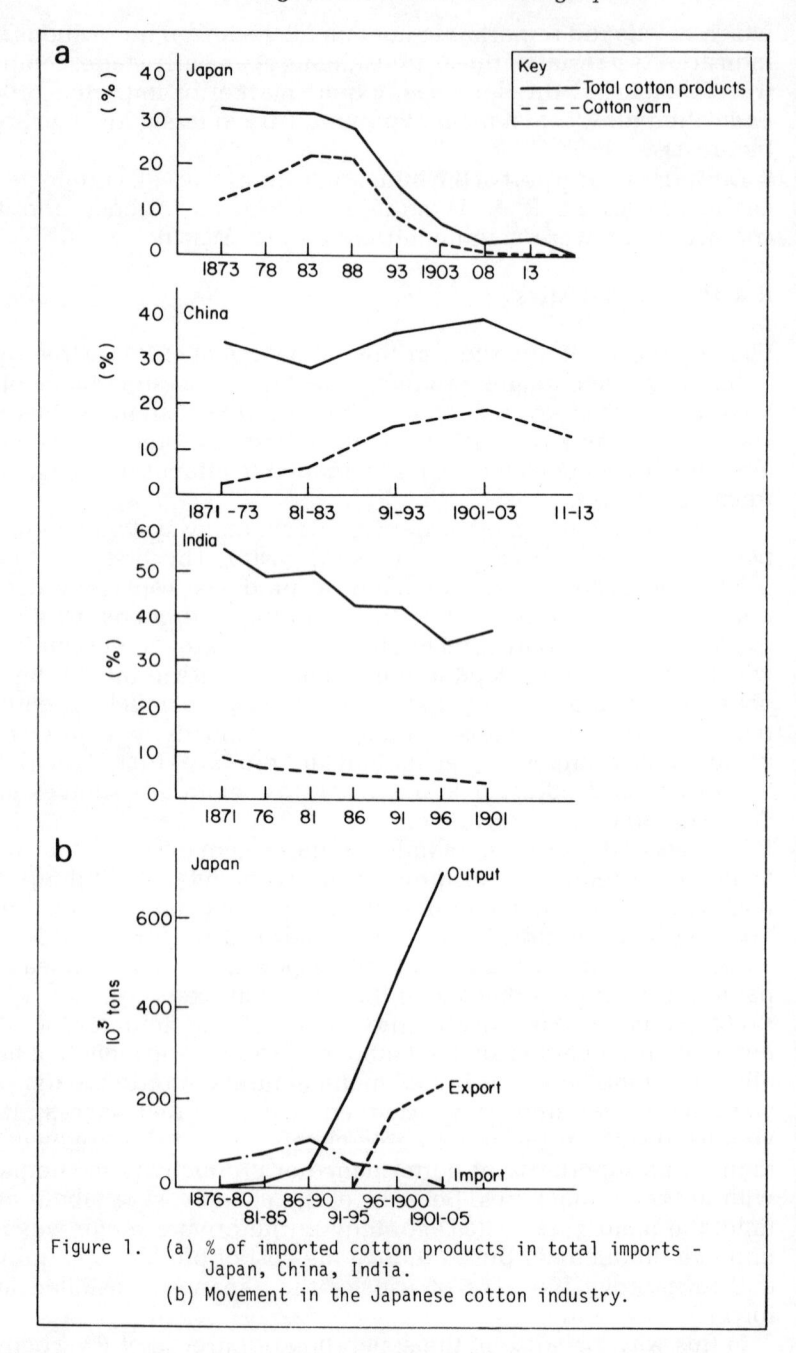

Figure 1. (a) % of imported cotton products in total imports - Japan, China, India.
(b) Movement in the Japanese cotton industry.

between the defence-related industries and commercial industry, industrial technology in Japan was formed with a dual structure. This dual structure of Japanese technology posed serious problems for the development of mass production technology, because mass production technology should be based on well-developed elements of technology, each being of similar quality.

Third, partly because of the dual structure of technology combined with the cheap cost of labour and partly because of limited and narrow markets, the Japanese machinery industry, including the defence industry, did not have a real enough incentive to develop mass production technology. This lack of mass production technology resulted in the low quality of Japanese machinery products.

Success factors in post-war industrialization

The next issue is how these characteristics of strategic policy in Japanese R & D before the second world war changed after the war, and how effectively the new strategy has been working in the post-war period.

Several years after the second world war, the Japanese economy and industry began the recovery process from the serious damage of the war. In particular, the outbreak of the Korean war in 1950 triggered the rapid growth in the Japanese economy (see Figures 2 and 3).

Looking back on the high growth of the Japanese economy, we can point to a variety of factors which contributed to economic growth; however, from the technological point of view, I point out three key factors for this great success in the Japanese economy.

Established production system

First, as mentioned above, Japan already had an established homemade production technology system in industry by the second world war, and therefore Japanese industry after the war could start a reindustrialization process on the basis of this system. For example, those industries such as the sewing machine, optical instrument, camera and shipbuilding industries reorganized the production technology and production facilities accumulated for the manufacture of military products, turned it to the manufacture of consumer products and industrial products, and in this way, in a short time, these industries had grown to be the leading export industries.

After the second world war, even in the high technology and new industries like electronics, synthetic fibres or petrochemicals, the accumulated knowledge from the old defence industry could help in evaluating the potential and importance of the new technologies to the government and private sectors.

The well known success story of Sony and other Japanese consumer

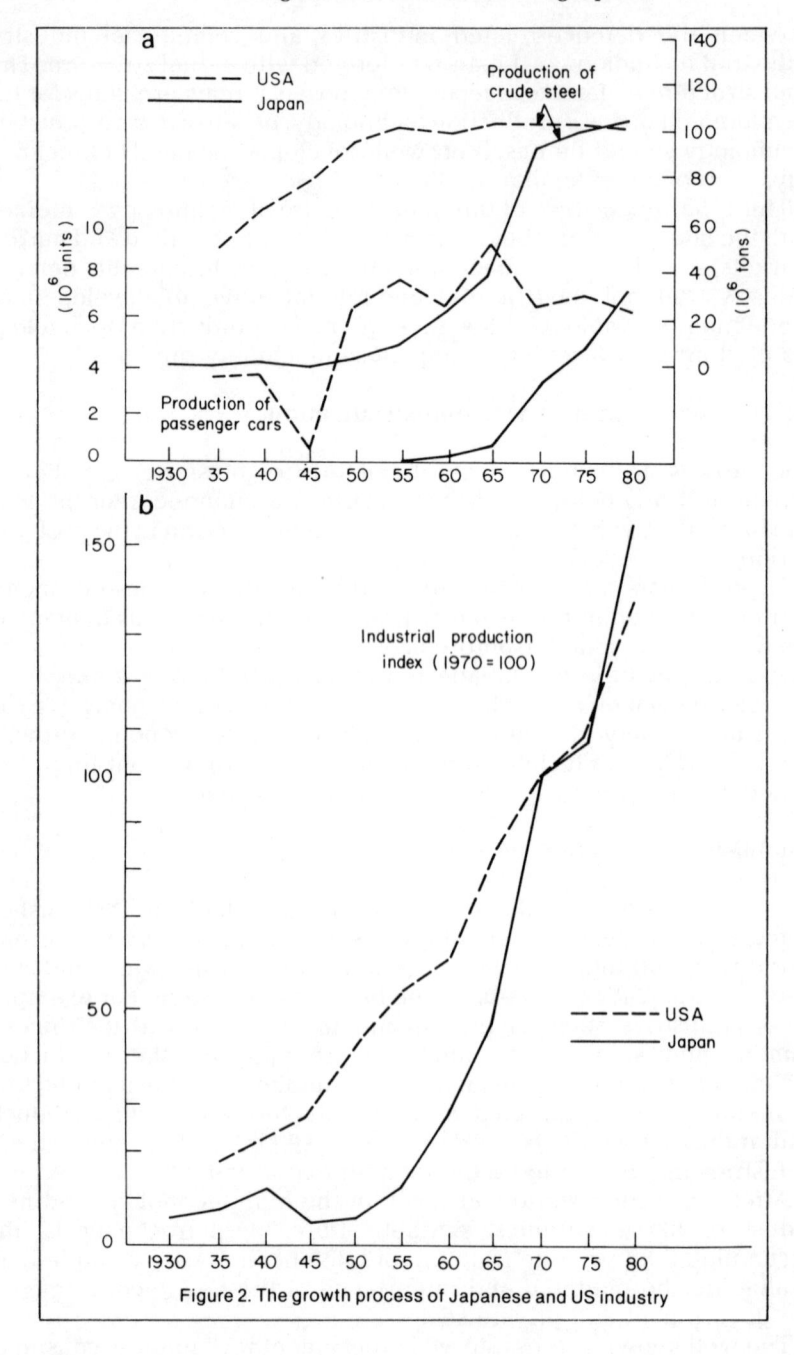

178   Design, innovation and long cycles

Figure 2. The growth process of Japanese and US industry

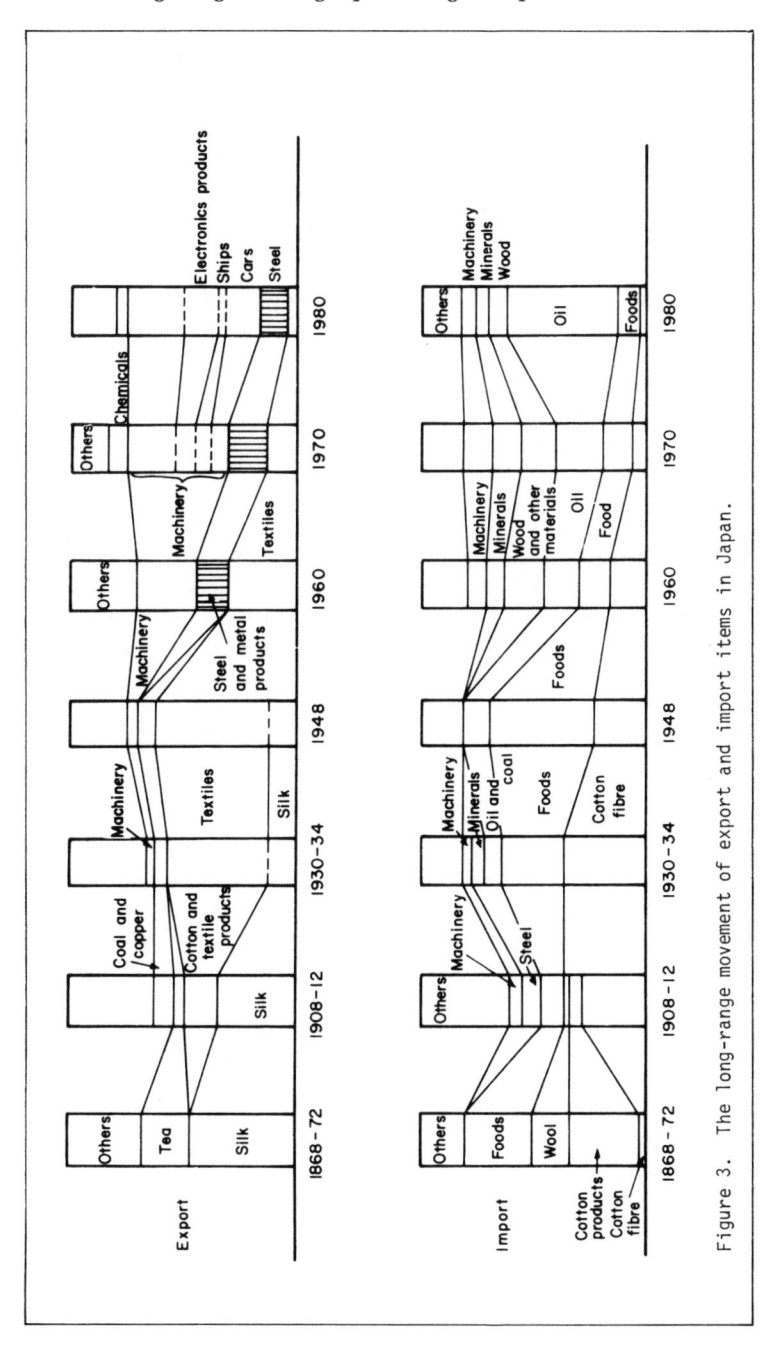

Long range strategic planning in Japanese R&D 179

Figure 3. The long-range movement of export and import items in Japan.

electronics companies is a good example of this. Because of the accumulated knowledge and engineering expertise in the defence industries released after the war, these companies were able to follow the innovations in electronics technology in the USA and apply that technology, developed initially for defence use, to Japanese consumer electronics products. In this way, the Japanese electronics industry developed transistor radios and small-sized televisions faster than US industry and was successful in creating a market for the new personal-use consumer electronics both in Japan and overseas. It was generally accepted that the key to the success of the Japanese consumer electronics industry, entering a matured market, lay in its success in the segmentation of a personal-use product market from the conventional home or family-use product market.

Thus, the homemade production technology systems established before the second world war could be used as an effective infrastructure and provided the basic form for the production technology system of post-war Japanese industry.

Incidentally, in any discussion of the threat now being posed by the growth of competitive industry in newly industrialized countries like South Korea, Brazil etc., we should bear in mind that these countries did not have the established homemade production technology system in the early stages that Japan had. Therefore, although they might at first glance show some similarities with Japan in their post-war industrialization process, there is in reality quite a wide gulf between the respective industrialization processes of Japan and that of newly industrialized countries.

Introduction of mass production

The second key factor depended on the success of introducing a mass production system, as mentioned above. The lack of a mass production system had been the weak point for Japanese industry in the past. After the war, when the Japanese defence industry had been completely destroyed and R & D activity was reorganized for commercial purposes, the US army general headquarters introduced the basic mass production method and the Japanese government and company management realized its importance for improvement in both productivity and quality of products. Therefore, they made the mastering of mass production methods their first priority. They invited over authorities on production engineering from the USA and organized many group tours to US factories to study the method.

Following the introduction of the mass production system from the USA, remarkable progress was made. In the 1960s, Japanese companies began to suffer from the shortage of low-cost young labour and they were obliged to employ part-time workers such as housewives and middle-aged people. Controlling and giving enough incentive to these unskilled and essentially low-quality workers was a big problem for mass production plants. In other words, they faced the problem of matching the low incentive workers to mass production technology.

Confronted with this problem, Japanese companies succeeded in providing incentives to the unskilled workers by organizing a variety of communal activities in each workshop, for example quality control circles for discussing and implementing improvements both in their jobs and in the quality of the products, and also many types of hobby circles. Every worker would participate on a basis of equality as a member of these circles. Because of the community spirit in each workshop, together with a lack of any clear task segregation, Japanese unskilled workers could feel a sense of family unity and showed loyalty to their community in the factory, and a kind of cooperative working system was enjoyed in each community. Japanese companies were thus able to organize these unskilled and semi-skilled workers, who received on-the-job training effectively and flexibly in their monotonous mass production workshops.

In the near future, because of the introduction of new microcomputer technology, the worker's job will become simpler and more monotonous, and the form of working will change from operative manual work to watching a computer display or a machine doing the work. Therefore, how to provide an effective incentive for this unskilled labour force will become crucial if adapting to the coming microchip revolution in advanced countries is to be achieved successfully. The existing Japanese management system and working system should provide some suggestions for the solution to this problem.

'Working textbook' for reindustrialization

The third key success factor was the existence in advanced countries of a working textbook for Japan's reindustrialization. Because of this, both the Japanese government, which has played the role of opinion leader and planner in the Japanese industrialization process, and private Japanese companies could reduce the political and business risk by gathering practical and useful information about new technology and new products markets from advanced countries. They themselves did not need to assess the risks and benefits of new technology. In addition to this advantage, it was relatively easy to form a consensus for growth both at the national and company levels.

The existence of the working textbook has a crucial influence over the form of Japanese industrial policy, industrial structure and corporate strategy, including R & D strategy. By referring to the working textbook in the period after the second world war, the Japanese government could design a systematic industrial policy to rebuild a full-ranging homemade production technology system based only on national companies; that is to say, selecting key industries whose technology had great potential for furthering the development of Japanese industry and providing them with a set of suitable political supports such as the low-cost industrial finance system, a protective trade policy, remission of taxes and effective subsidies for R & D in new key technology.

For example, from the late 1950s the Japanese government selected the electronics industry as one of the key industries of the future. By the latee 1960s, Japan was obliged to reduce its tariff barriers even in the cradle industries like computers and integrated circuits (ICs), and under these circumstances the Japanese government organized a series of suitable projects to strengthen key electronics technology with the clear target of catching up with IBM technology. Needless to say, the most important success factor in R & D activity is setting a clear target with the right timing, and with the working textbook Japan had plenty of information with which to achieve that effectively at relatively little expense.

In the successful development of very large scale integration (VLSI) technology one can see the clearly apparent function of the working textbook. VLSI technology consists of more than 20 innovative elements of technology such as sophisticated metal-refining technology in IC material industries, special photo-resin technology for laser beams, ultra-pure air-cleaning technology in the construction engineering industry, optical instruments, fine mechanical technology and measuring instrument technology, and so on. These elements of technology, which were mainly supplied by the non-electronics industries, had to be combined with sophisticated IC technology from the electronics industry to form the VLSI technology system for the achievement of what was a national project.

In a leader country like the USA, new technology would be developed and formed naturally through the market mechanism, and small companies like venture businesses would play an important role in each basic technology field. But in a follower country, elementary technology has to be organized artificially to some extent throughout a project.

In the development process of VLSI, I do not think the government subsidy of about $40 million for the project was the most important factor for success - the critical factors were the setting up of a proper target and the organization of such a variety of companies into one project. Needless to say, Japanese industry found a suitable working textbook in the USA and it was possible to form a wide-ranging consensus for this national project.

Influence on structure and strategy

I now discuss the influence of the 'working textbook' on Japanese industrial structure and corporate strategy. Owing to the existence of this working textbook, key Japanese industries have a competitive, oligopolistic structure and share a hierarchy of technology and labour between firms, according to their business scale. That is to say, the existence of working textbooks has meant that, because certain critical technology was imported, every established large company was able to take the business opportunities simultaneously. In addition, government control meant that only big companies could qualify to be licensees for foreign licensers, and only they were able to make the large-sized investment for new products that this entailed.

Hence key Japanese industries like automobiles, petrochemicals, oil refining, iron and steel, computers, ICs and other electronics industries and consumer electrical goods industries have more than five or six main suppliers. In these industries, there is no one company which has a lion's share of the market and plays the role of price leader. Therefore, the key companies in each industry are forced to compete keenly for market share and none is able to earn innovators' monopoly profit and market share.

In addition to the technological factors, most big companies are arranged in what are termed 'Zaibatsu' groups, which usually centre around large banks. In the past high-growth age, each main bank used to act as money supplier to the big company of its group. Even if a large company failed in its business, their main bank would provide money and help rebuild companies. Therefore, the mechanism of natural selection hs not worked effectively in each of the key industries, and the keenly competitive market circumstances have remained without any significant change.

In these market circumstances, it was inevitable that Japanese management became aware of the importance of product discrimination as a production and marketing strategy, because each of their competitors has the same basic imported product technology and they compete with their rivals by means of product discrimination and price reduction based on the scale economy of production volume. Adding to this factor, as long as the working textbook was there, it was more effective and economic to import promising technology from abroad than to develop it themselves. Thus Japanese companies have poured enormous business resources into developments for product improvement, in quality and design, and process innovation as an R&D strategy, rather than developing basic technology or original products. Needless to say, technology for the improvement of products and production processes depends on the more practical and day-to-day business activities in the production division, rather than on the purely scientific activities of the laboratory. Therefore, as a result of this corporate strategy, Japanese companies' production departments, and not R&D departments, have accumulated a significant amount of business resources and gained a kind of authority in their companies. In this way, Japanese industries have been able to establish very competitive and efficient production capacity.

This corporate strategy may be quite different from those of Europe or America, especially in the USA where production divisions were seen as the most tedious divisions, controlled by computer systems engineers and production engineers, and whose managers were felt to be the last people for promotion to top management.

It is clear that the characteristics of the Japanese types of effective R & D activity and corporate behaviour are mainly due to the unique Japanese industrial structure and corporate strategy.

We now look at the disadvantages which result from the existence of a working textbook. As long as Japanese companies import the seeds of new business from abroad, they are doomed to struggle with strong overseas competitors in international markets as well as in domestic markets. Therefore, trade friction between the licensee country and the licenser country would seem to be inevitable if the licensee makes a success of the business with imported technology.

Another disadvantage of a working textbook is that because of Japan's policy of selectively picking up the potential key technology from each industrial stage, from the materials industry to the finished goods industries, and assembling that technology for use in Japan, the established Japanese industrial structure seems designed to deny other advanced countries the benefits of the international division of labour. An accusation levelled at Japanese industrial structure has been that Japanese industrial policy acts against international free trade rules.

Recent movement in Japanese R & D activity

Following the oil crisis, the Japanese economy and industry faced a critical turning point, and the effectiveness of the working textbook has been diminishing. For further development of the economy, Japanese industry and government should adjust their industrial policy and corporate strategy to these new circumstances to some extent.

Japanese industry must switch from its basic position of follower to a position of leader in the world. Such a switch is not an easy task because it entails a reformation of industrial structure and corporate structure.

However, there are a few indications that this switch is beginning to take place. Companies in Japan have already started putting a strong emphasis on R & D activities to create original products and technology. To complement the R & D activity of private companies, the Japanese government has started several new national R & D projects like the well-known innovative fifth generation computer technology, thermonuclear fusion technology, biochemical technology, etc. On a short-term basis, I think that because of the sophisticated Japanese types of improvement and application technology, combined with Japanese-style management and working systems, Japanese industry has enough potential to become one of the leading countries of today's microchip revolution and to produce new products and develop new markets.

However, there are many obstacles to wide diffusion of recent technical innovations, such as the high unemployment in advanced countries, continuous trade friction not only with established products but also in new innovative product sectors, the huge budget deficit mainly in the government sector, and the high real interest rates. In facing these problems, we cannot take an optimistic view of the future growth of both the world economy and the Japanese economy. Especially in Japan, with the loss of the textbook, each company will have to make its decisions with reference to its own criteria of profit and loss, and I am afraid that Japanese companies are becoming cautious about investing their business resources under uncertain business circumstances.

10 Marx's crisis theory today

*Francois Chesnais**

Today I'm going to say certain things on which I do not seek a united front with anyone but will invite discussion later.(1) Carlota Perez says at one point in her paper that capitalist economy and society moves from crisis to crisis. This is a sentence Marx could have used, and in fact I think it is almost a literal quotation.(2) Before examining the features of the present structural crisis, I think it may be useful to say a few words on what Marx's position might have been on the issues that we are discussing.

In Marx you have a theory of crisis rather than of cycles. Marx noted the tendency, in the middle of the 19th century, towards a ten-year recurrence in crises. He discussed this briefly and formulated the hypothesis that this was probably due to the bunching of capital investment in combination with the current life-cycle or rotation periods of capital equipment at that time.(3) But he did not push this very far, because his interest was less in the cyclical aspect than in the much more important aspect that capitalist society tended to move from crisis to crisis.

In his work, crisis has two dimensions: one the economic, the other structural.(4) Turning first to the economic aspect of crisis, it is always related to the fall in the rate of profit, to the point where profit is insufficient to warrant further accumulation. Marx discusses overaccumulation only relative to the rate of profit. He is careful to say that there is never such a thing as an absolute overaccumulation of capital, but only in relation to the profit rate.(5) On the one hand he sees a central contradiction of capitalism in the fact that the relations of production are also the relations of income distribution. At the same time that capitalism is tending to drive forward the development of productive forces, it is doing it on the basis of a tendency to keep wages down. This is one of the central contradictions of capitalist development.

At the same time he was definitely not an underconsumptionist.(6) He said that insufficient effective demand in the form of consumption is already an expression of the fact that accumulation and the investment process have slowed down and that less wages are being paid out, and that it is subordinate to what lies on the side of accumulation and on the side of the drop of the rate of profit. Behind this drop in the rate of profit lies the tendency towards the rise in the organic composition of capital.(7) This is the central thrust of everything he has to say about the way a long swing, a long wave of capital accumulation is first fed. Then it begins to peter out as the movement towards the rise in the organic composition of capital asserts itself. This happens in spite of the factors which act as counter-tendencies.

It is there that we have a very interesting development in relation to technology, which Nathan Rosenberg pointed out in a very useful article he wrote for 'Monthly Review' about ten years ago on Marx as a student of technology.(8) He noted that some of the most interesting things Marx has to say about technology are related to the way in which the process of capital accumulation reacts to what is simply a tendency toward the falling rate of profit, by fighting it. It fights it in particular through technology as a means of raising the rate of exploitation and bringing the tendency down, or fighting rises in the value of the constant parts of capital - that is, fighting rises in energy, material and capital costs, all these are part of this process. So part of what has to be said about technology in a Marxist approach is that it is an integral part of this process of: accumulation, the rise in the organic composition, the tendency for the rate of profit to fall, and the way the system fights it over a long period of time.

The economic function of crisis is a very important aspect of the discussion. In Marx the economic function of crises, made very clear by Mattick,(9) is that of ensuring the destruction of a sufficient amount of capital, whether in the form of physical capital or of goods and stocks, to create the conditions allowing the expected profit rate to return to a level sufficient to allow accumulation to begin again. This corresponds to what Schumpeter calls 'creative destruction'.(10) One of the functions of crisis is to ensure the cleaning up of the conditions of the rate of profit along with the simple clearing of the decks for a new wave of investment.

The structural dimension of crisis begins with the way in which capitalism does this. Capitalism achieves this clearing of the decks in a way in which the social costs, and the resultant growing social resistance, increase from crisis to crisis, because capital develops through crisis and extended periods of unemployment and all that goes with the process of devalorization of capital. This is why Marx was led to say that crises are two things. They are a necessary function within capitalist development, for capitalist development cannot take place without crises. But at the same time they are an expression of the contradiction between the productive forces which are potentially too big, too strong for these capitalist property relations.(11) Right from 1847 he says that crises express the fact that capitalist property relations are ultimately too narrow to contain productive forces and resources and to manage them satisfactorily from society's point of view (or at least from the working classes' point of view).(12)

What Schumpeter has to say about Marx is in Marx: that capitalism creates through its success the conditions where other sets of property relations, other forms of social management, become both possible and necessary - necessary in the sense that if they are not adopted then only more catastrophic expressions of crisis conditions can develop.(13) The structural dimensions of the crisis may be seen to increase over time. Marx noted this himself when he spoke about the 1876 crisis.(14) He noted that a certain number of features in development meant that the crisis would be both deeper and more universal in its effects. This was an expression of this movement from crisis to crisis.

Both in the 'Grundrisse'(15) and in volume 3 of 'Capital',(16) Marx refers to a whole set of what Perez calls 'social and institutional innovations of capitalism', of which he was very centrally aware. He argued that capitalism was finding ways to push back the barriers limiting its development, which were only creating the conditions for an even stronger contradiction between productive forces and relations. This would develop within these larger barriers and would persist because of the basic contradiction of property relations.(17)

In this way capitalism is moving from crisis to crisis. By the inherent means of its success in solving the previous crisis conditions and of finding some of the new institutional responses to the process of socialization and internationalization of productive forces, within the limits of private property relations, it only creates over time the conditions for ever bigger crises and more potent contradictions. In the same way that we talk about economic trends, Marx saw within capitalism a two-fold socialization: of internationalized productive forces and their corresponding economic or social relations. This theme of capitalism's development of increasingly dense social relationships is very strong in Marx,(18) and becomes central to Schumpeter, who recognizes his debt to Marx in this respect.

Within this context, undoubtedly Marx sees technology as a potent force which, in each phase of accumulation, both feeds the accumulation process and is a force which increases the socialization and internationalization of production. The letters Marx wrote after going around the 1851 exhibition in London, looking at the technology, express the theme that these inventions are wonderful:(19) they establish the basis for managing society in another way; they are too strong, too potent to be held within the limits, within the bonds of capitalist property relations. This is the basis of one of his approaches to technology.

So moving on to the second part of my argument, we can say that crises are phases reflecting the forms evolved by capitalism for managing economic and social life on the basis of private property relations, and in particular the response to the need to clean up the way in which it actually manages 'creative destruction'. Crises are the phases where the way in which it does this calls into question the very basis of these property relations and raises (both at the conceptual level and the concrete level) the issues of whether there are better ways of managing social and economic affairs.

Now this is basically what a crisis is, for a Marxist. It is the moment in which the system itself, by the inherent logic of its working, calls for open dicussion of this, with millions of unemployed for emphasis. Are there not better ways than those under which capitalism has developed of managing these productive forces and this socialization of social life and this internationalization of economic life? The basic property relations and most of the ideology on which this system is based contradict thse developments. Much of this is in Schumpeter as well. Schumpeter's pessimism comes from the fact that he too felt that capitalism was doing this structurally by means of its very success.

Look at the 1929 crash and the subsequent world crisis, the Nazism which followed it as an initial response, and the second world war which decided, temporarily, a host of questions. This was a major structural crisis of the type which Freeman was thinking of when he asked me to comment on this issue, because this period and those events did three things.

Here I will be brief because Perez also deals with this. Firstly, this series of events called into question the existence of capitalism. Secondly, the events prompted a series of social and institutional innovations by capitalist forces as a response to this challenge to its very existence. Thirdly, the whole of these events cleared the way for a long period of accumulation through the double process of protracted disinvestment and a very important ageing of the capital stock that remained. This was accompanied by the effects of World War II in the massive destruction of physical capital and productive forces and, of course, men. But it did so within certain conditions which can never be forgotten. The existence of capitalism was not only called into question theoretically: the Chinese revolution and a certain number of other things were material results. The structural crisis of the thirties and forties did end with change in the nature of property relations over about a third of the globe. This is the sort of thing people forget or put at the back of their mind, but in any discussion of that structural crisis one must not forget it.

The other point is that even where capitalism ultimately managed to regain strength it did this on the basis of a set of social and institutional innovations which involved a very deep break with the previous situation and practice. There had been some experiments, even by Bismarck, in attempting to get some kind of social consensus with organized unions and also give the working class some improvement in conditions with regard to health, social insurance etc. However, the big break came at the end of the 1930s and after the second world war, where there was a complete break with previous practice and a new set of constraints - which capitalism accepts because it is the only way to start accumulation again. But as time moves on, these new conditions are felt more and more as constraints. In the long phase of expansion and accumulation which follows, we have three interesting vital things which are central to the nature of the crisis today.

The first is that we have an advance towards the socialization of social life within the limits of private property relations without precedent in history. Now this is the objective product of the ways economies have evolved: the concentration of bigger more socially organized ways of doing things, of communication, of transport, of dissemination of information, of everything. All this is moving objectively towards the socialization of economic and social life, on which I think there is no going back.

Secondly, this socialization is the result of the struggle of political forces, incorporated into a series of institutions and also into a series of expectations - a whole culture, a whole way of seeing. Generations have now been educated in a certain way of understanding a whole set of social phenomena and a whole set of things which they think they are entitled to expect from organized society. Now much of what I read on 'social rigidities' is, for me, a discussion of all this and of much that is in Schumpeter. He said that there may be a fourth Kondratieff but if there is, its very success is going to push all the tendencies I have already noted even further. The results of this are perceived today as social rigidities, rigidities of all sorts. This is harking back to a previous capitalism which did not have these rigidities, instead of trying to get to grips with the fact that socialization has become as dense and intense as it has.

The second phenomenon is internationalization of economic life. This is the phase which was ushered in by the end of the Second World War, by Bretton Woods, by the United Nations Organization, by the Marshall Plan, by OECD, by the Common Market, by all these institutional arrangements which permit capital to flow internationally in a way which had never been possible in the previous part of the 20th century. All this has led to an internationalization of economic life and of productive forces in a way which is unprecedented. Now I do not need to insist on that because as soon as you start thinking about it, it is so obvious and it can be ilustrated in so many ways.

The multi-national corporation is the capitalist response to the internationalization of productive forces. It attempts to organize this within the limits of the company and to found the profits and the growth of the company on the exploitation of the international dimension of productive forces. It can only do so within limits, but it is an expression (though not the only expression) of this tendency. When productive forces outrun the limits of national boundaries in a way which has no precedent in history, this moulds economic policies.

The only way in which governments can act today is by helping their firms to be more aggressive on world markets; retreats into protectionism are practically utopian.

Now let's move to technology. In my approach, what happened in this post-1945 upsurge is the following. This phase of accumulation involved a series of technologies within a context in which the social and institutional innovations permitted their incorporation. A mode of accumulation based on the appropriation of unpaid labour through the techniques of Fordism and Taylorism was necessary, but in conditions where sufficiently large numbers of workers could be incorporated into this process. So a wide enough market had to be pushed further than it had ever been pushed before. But at the same time this phase, as Perez points out in a different framework, has given rise to a technology about which Marx wrote two paragraphs in the 'Grundrisse': the type of technology which he said would make the appropriation of unpaid labour puny and irrelevant.[20] Microelectronics, microprocessor technology and information technology are the technologies which make previous forms (or in particular those forms which have been pushed furthest in this last phase up to 1970) look puny now.

The central question which this raises is the following: the whole of capitalist property and social relations of production, the whole of capitalist economic relations have been based on the existence of wage labour. Capitalism is the social system which brought the peasant artisans into the working force, destroying their previous way of life, developed a huge working class and geared all techniques to the appropriation of unpaid labour on as massive a scale as possible. And now in its development it has thrown up a technology which makes this both irrelevant and no longer possible, and this is the real challenge. It initially appears as simply a challenge to traditional labour unions which have been educated in the fight to get labour paid a decent a price on the market and to fight for part of the unpaid labour. The whole process looks as if it is hitting them, but ultimately it hits capitalism itself as an instrument, as a set of relations, even harder, because capitalism has drawn this working class into existence. It has created a form of labour which goes far beyond the working class proper, which goes into services etc.; it has created a society where a man or a woman is only someone to the extent that they can sell their labour. One is nobody in this society, and this was discussed deeply in the Latin American debate on marginalization - one is nothing if one cannot actually get into the economic process. Now what can capitalism do about a technology which now makes all this absolutely irrelevant and pointless, which requires work to be organized, shared and based on completely different criteria.

This is the point at which what I have to say today joins with Perez. This is the central issue, in the face of which there are going to be a series of reactions or a series of replies, pseudo-replies, and non-replies. I think that in view of this we are going to live 20, 30, 40 years of chaos, which is going to be both economic chaos and social chaos. What is required are is the social and institutional innovations which can actually cope with what microelectronics technology in all its ramifications means. Now I believe it just cannot be done on the basis of private property relations and that this is something to which bourgeois forces have ultimately no reply.

Whatever face socialism may have in the 20th century (in the questions, in the methodology, in the way Marx poses the question), which I have tried to relate to these current issues, it still addresses the question of the further development of society. For the relations of production to encompass this socialization, this internationalization and these new quite revolutionary technologies, requires this complete break with the whole way accumulation has been previously structured.

Notes

* Prepared for publication from transcript, by Steven Little and Tim Martin.
1. A paper is available: 'New features of the international economy and technology compared with earlier structural crises', which I wrote for an earlier conference. It is a paper in which I was seeking a united front with people like Nelson in a rather hostile environment, and so it should be read in that context. Now all this might seem to you rather queer, coming from someone who works at OECD. But at OECD I sell my labour-time. I'm one of the rare intellectuals who accepts this, that he actually sells his labour-time. I think I do a decent job in OECD but I'm just selling my labour-time there, and this allows me to keep my own ideas for myself. But this is a sufficiently propitious environment in which to put things as I see them on the issues we are discussing here.
2. K. Marx and F. Engels, Communist Manifesto, in 'Selected Works' (Lawrence & Wishart, 1968) pp.40-41; and Wage Labour and Capital, ibid., p.93.
3. K.Marx, 'Capital', (Lawrence & Wishart): on cycles and crises: vol.1, 1954, p.29; on ten-year cycle, life cycle of fixed capital: vol.1, pp.593, 596, vol.2, 1956, pp.188-9, vol.3, 1959, pp.489-93; on crises and credit: Marx & Engels, 'Letters on 'Capital ' (New Park, 1983), 2 March 1858 and 5 March 1858; and P. Mattick, 1981, pp.73-76.
4. The best statement I know, in English, of Marx's theory of crisis, is Paul Mattick's 'Economic Crisis and Crisis Theory' which was translated by Merlin Press in 1981.
5. K. Marx, 'Capital', vol.3, op.cit., pp.250-59.
6. See Marx's criticisms of Sismondi in 'Theories of Surplus Value' (Lawrence & Wishart, 1971) vol.III, pp.55-56.
7. K. Marx, on the organic composition of capital and the rate of profit: 'Capital', vol.1, op.cit., ch.XXV; and vol.3, op.cit., part III.
8. N. Rosenberg, Marx as a Student of Technology, in 'Inside the Black Box: Technology and Economics', Cambridge University Press, 1982.
9. P. Mattick, op.cit., on the economic function of crisis, pp.104, 135-6.
10. J.A. Schumpeter, 'Capitalism Socialism and Democracy', Routledge Kegan Paul, 1976, ch.VII, p.81.
11. K.Marx, 'Capital', vol.3, op.cit., pp.249-50.
12. K. Marx, Wage Labour and Capital, op.cit., p.93.
13. J.A. Schumpeter, op.cit., p.162.
14. Marx to Danielson, 10 April 1849, in 'Marx-Engels Selected Correspondence', (Progress, Moscow, 1975) p.296.
15. K. Marx, 'Grundrisse', (Penguin, Harmondsworth, 1973); see, for example, Marx's discussion of competition and monopoly, pp.649-52.
16. K. Marx, 'Capital', vol.3, op.cit., pp.249-50.
17. P. Mattick, op.cit., pp.55-57.
18. K.Marx, 'Grundrisse', op.cit., pp.706, 712.
19. Marx to Engels, 13 October 1851, in Marx-Engels 'Letters on Capital', New Park, 1983.
20. K. Marx, 'Grundrisse', op.cit., pp.704-6.

11 Long waves and the international diffusion of the automated labour process: the role of the semi-industrialized countries

Leonel Corona

The unequal development of capitalism in time, spatial distribution and content, and regularities in its development under the form of competition, and the evolution of technical change in the labour process, are the points of departure of this paper in trying to understand the new trends of capitalism in the semi-industrialized countries (SICs). 'Time' is considered by way of historical crises of capitalism that define the long waves of expansion and contraction; 'extension' by way of looking at some of the historical features of the international expansion of capitalism in which the so-called developing countries have played a dominant role during the phase of imperialism; while 'content' refers to the level of the forces of production that have been adverse to the SICs, where considerable sectors of the population have not yet been able to satisfy their basic needs.

How this uneven development will continue depends on changes in the regularities of capitalism: deeply entrenched state monopoly capitalism is looking for enlargement and strengthening of regions of influence, and polarizing the dominant contradiction between socialist and capitalist countries. Arguing that diffusion of automation in industry and services is the new technological basis for increasing productivity, the framework for labour value is changing. The emergence of these fundamental changes is a matter of state regulation and institutional change.

SICs are in the present crisis the weak link of the international financial system. The question is: How will they continue to participate in the international valorization of capital both by borrowing money and increasing foreign investment, thus 'opening' the economic space for a larger industrialization wave based on automated labour in a possible future upswing of capitalism.

Evolution of the labour process of production (the logic-historic approach)

The labour process of production is the material relationship between labour-power, the means of production and the object of work.

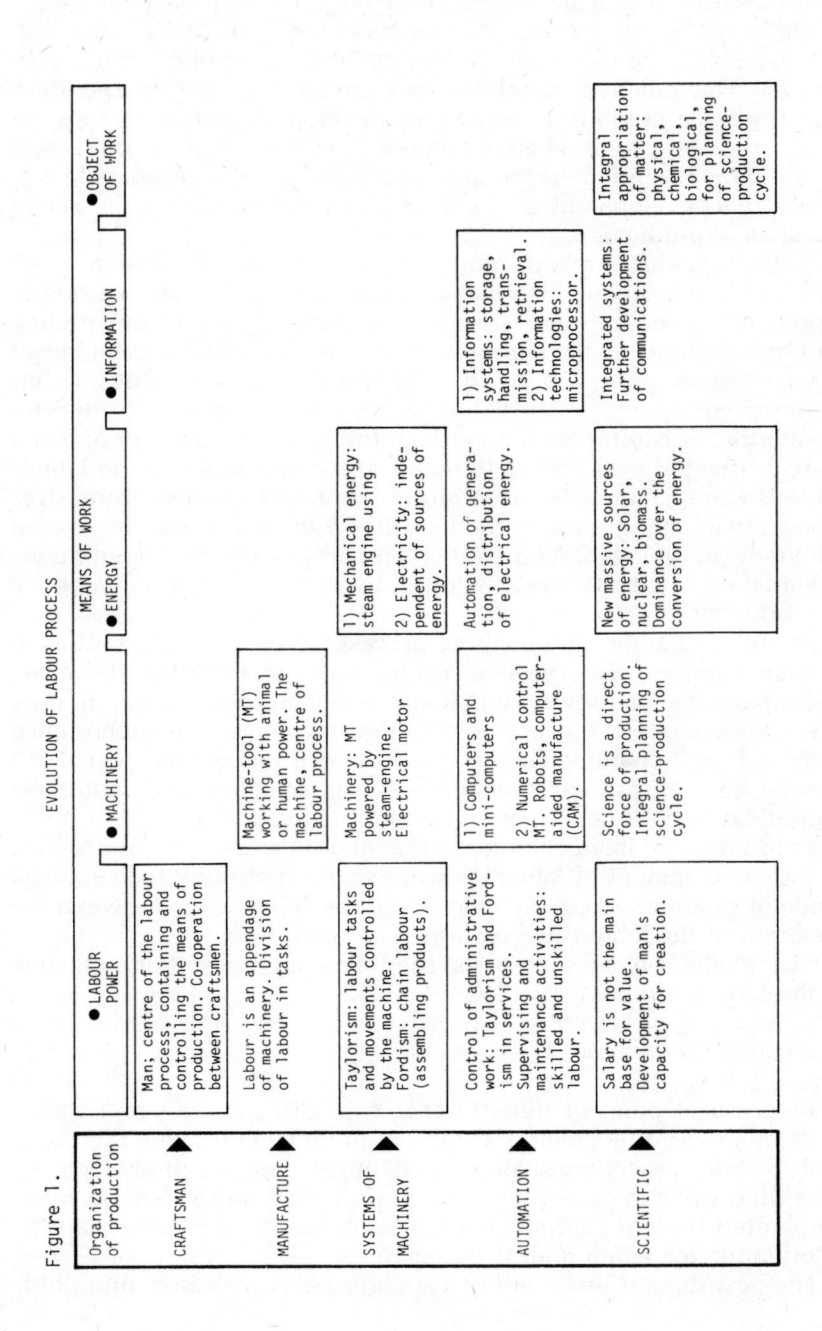

Figure 1.

EVOLUTION OF LABOUR PROCESS

Organization of production	LABOUR POWER	MACHINERY	MEANS OF WORK — ENERGY	INFORMATION	OBJECT OF WORK
CRAFTSMAN	Man; centre of the labour process, containing and controlling the means of production. Co-operation between craftsmen.				
MANUFACTURE	Labour is an appendage of machinery. Division of labour in tasks.	Machine-tool (MT) working with animal or human power. The machine, centre of labour process.			
SYSTEMS OF MACHINERY	Taylorism: labour tasks and movements controlled by the machine. Fordism: chain labour (assembling products).	Machinery: MT powered by steam-engine. Electrical motor	1) Mechanical energy: steam engine using 2) Electricity: independent of sources of energy.		
AUTOMATION	Control of administrative work: Taylorism and Fordism in services. Supervising and maintenance activities: skilled and unskilled labour.	1) Computers and mini-computers 2) Numerical control MT. Robots, computer-aided manufacture (CAM).	Automation of generation, distribution of electrical energy.	1) Information systems: storage, handling, transmission, retrieval. 2) Information technologies: microprocessor.	
SCIENTIFIC	Salary is not the main base for value. Development of man's capacity for creation.	Science is a direct force of production. Integral planning of science-production cycle.	New massive sources of energy: Solar, nuclear, biomass. Dominance over the conversion of energy.	Integrated systems: Further development of communications.	Integral appropriation of matter: physical, chemical, biological, for planning of science-production cycle.

Changes in this relationship are quantitatively related to productivity in the sense of increasing or decreasing output in relation to the inputs of the labour process elements. But it is also a political relationship in terms of the control (or the ownership) of the elements in the labour process. This means that productivity changes are not independent of the social conditions in which they develop. Therefore, the way in which productivity is changed depends upon the logic of capital accumulation, embodying a specific fundamental contradiction between the development of the forces of production and the social relations of production.

In the capitalist mode of production man is historically converted into a 'factor of production', that is, a means of production not simply producing goods to satisfy needs, but primarily producing profits. Thus the evolution of the labour process must be regarded as technical change being propelled by the inherent laws of capitalism, but nevertheless based on available techniques. On one side, the accumulation of capital tends to reduce the value of labour in order to increase the surplus, and on the other the organization of the labour process changes in order to retain or increase capital's control over labour; both changes impel the use of new technologies. The overall tendency in the evolution of the labour process is a continuous devaluation of labour-power with an increase in the social capacity of production.

The basic changes in the labour process are those in which the role of man changes. This is a necessary point of departure because, resting on the increased complexity of technology, labour is only considered as one of the factors of production and is losing importance with technical change. This is the case with innovations which are viewed just as products or processes (machinery and materials) regardless of the role of man in the production process.

To identify the main changes in the labour process we have to look at the development of labour-power and its control by the capitalist mode of production, i.e. the changes in the relationship between the elements of the labour process and its organization.(1)

Historically it is possible to list the following types of organization of the labour process:

Craftsman production

The technical point of departure of capitalism is craft production. Here, man has direct overall control of production through the use of tools. Labour-power participates in all the elements of production, i.e. providing skills augmented by tools, providing muscle energy complemented by other primitive sources of energy, and providing the information for relating all the elements.

The new dimension added by capitalism is cooperation, brought in by the gathering of many craftsmen, and the potential division of labour through the specialization of tasks. At this stage the worker has real control of the labour process despite the formal appropriation of it by capital. The greatest possible technical change under this organization might be the specialization and perfection of tools.

Manufacturing production

Once the cooperation between craftsmen is settled through the division of labour, the conditions for the introduction of machine tools arise. This begins the real appropriation of the labour process by capital, because from this point on the machine will provide the framework to rule the process of production on its behalf.

The machine-tool not only incorporates the technical knowledge of the craftsman but begins the systematization and standardization of the labour process. Although the machine-tool is the determinant element in the industrial revolution, nevertheless it is the steam engine which is the dominant technological aspect.(2)

The type of knowledge which must be related to manufacturing production is applied mechanics, which makes possible both the design and operation of machine-tools and also (no less important) the interrelationship of the machines.

System of machinery

The developments of the means of production up to the point where different machine-tools could be interrelated is in fact done through the introduction of new available sources of energy. So, the labour process is embedded in a system of machinery, in which worker participation is dictated by the logic of the machine. The tendency to minimise labour costs (or from the other point of view increase labour productivity) is strongly developed, through cooperation between machines where labour is just an appendage. The division of labour into routine tasks matching machine operations extends to what is known as 'Taylorism' which is the description of labour task organization as part of a system of machinery. 'Fordism' or chain production is a related system where the end product is assembled from many parts (as is the automobile).

The energy required for this industrial revolution, in which production is based on cooperation between machines fed by one source of energy, has developed in two stages: mechanical energy supplied by the steam engine, and electricity. Electricity, or 'secondary energy', allows the use of a variety of primary sources of energy (coal, oil, hydro-electric, nuclear, biomass, solar), and thus creates a modern system of energy supply.

The system of production dependent on electricity is related to the electrical motor used to power the system, since the same technical principles are inversely applied to the dynamo or electricity generator.

New branches of production are created with the development of the industrial revolution. Besides the expansion of capital, the key for its accumulation logic is: first, the industrial production of machinery, and second, the massive production of goods. This means that the accumulation of capital can devalue labour-power through increasing productivity in those sectors that produce goods for the workers and in those that produce machines, therefore generating counter-tendencies to the falling rate of profit caused by the increase of constant capital (embodied in the mass of machinery, raw materials and facilities) relative to the variable capital (used to purchase labour-power).

Automation

The tendency to diminish labour participation in the process of production creates the opposite effect, because with the introduction of information technology the role of labour changes qualitatively from being a part of the machinery-mechanism to supervising and maintenance activities. There is thus an inversion of those tendencies because labour, as a collective-worker, operates an entire system of production through the introduction of microprocessors in different parts of the process, utilizing computers for the integral control of production.

Before coming to the industrial application of information technology, it is necessary to consider its application to administrative activities. The introduction of information technology began with the handling of huge amounts of calculations and data through the use of computers, and therefore the role of information technology in the service sectors should be seen in the perspective of capital accumulation.

Information technology can be considered as coming into existence with the invention of the computer. Its initial applications were in the administration, handling, storage and retrieval of information, but from the labour process point of view computers are also a tool for controlling, standardizing and minimizing workers' participation. Therefore computers give a technological basis for creating systems of information in which they are the central part of the administrative labour process, in the same way that machine tools are the central part of the systems of industrial production.

The key innovations in information technology are the transistor and microprocessor. In 1950 the former replaced the vacuum tube, allowing the development of integrated circuits. This electronic miniaturization progress produced the microprocessors that are now the key element in the automation of industrial production. The technical feasibility of auto-regulation in the production process is shown by the microprocessor because it can work on line, it is reliable and can control simple physical variables, but mainly because of its increasing cheapness.

Automation gives more integration to the process of production and more flexibility. For instance, in the case of transfer-machine-tools the handling of materials is combined in one machine with the tooling operations, using automated control devices. Also, in the universal-machine-tool the entire tooling operation is performed under the control of a programmed tape or magnetic disc without moving the object.

With automation, labour skills become polarized into two groups: one that needs special and multiple qualifications for the maintenance and programming of the operative alternatives of the machinery, and another, larger, which supervises the process and is responsible for minor tasks of the operation (lectures, cleaning, feeding raw materials).

The computers and microprocessors that allow automation are basic innovations in the sense that gives rise to entirely new industries(3) with an impact on nearly all economic activities. Moreover, the information technologies qualitatively change all the elements of the labour process: the labour-power is reduced quantitatively and is polarized between those who are skilled and those who participate in marginal activities; energy is becoming a large integrated system of production, transmission and distribution by the incorporation of automated control equipment; production of information technologies is becoming, partially, an automated 'filiere' from semi-conductors to the production of instruments, machines and equipment in which they are incorporated.

Furthermore, new conditions in the automated process of production can give rise to new forms of organization. Automation alters not only the production process but also the product, in that both are designed together, resulting in a simplification of the parts assembled in a flexible organization of production, giving alternative characteristics to the outputs. This means an inherent requirement for better knowledge of the physical, biological and chemical characteristics of the materials used to make the product. The internal evolution of the labour process requires the incorporation of science, when the design, planning and set-up integrate both the means of production and the products. Science could become a more important element in the labour process of production through the development of scientific organization of production.

Scientific organization of production

There is an intrinsic need for deeper knowledge of the physical, chemical and biological characteristics of materials when automation generates an integrated production-product process. This technical demand for knowledge constitutes a new characteristic of science in the sense of becoming a direct productive force of social production. This concept means that stronger links between science and technology are arising from what is nowadays called the scientific and technological system. The relationship between science and technology can be divided into a first stage, when science broadly follows technological advances, to a new stage, in which the development of technology depends upon the advance of scientific knowledge. This is now the case for many of the more technology-intensive sectors[4] (biotechnology, aeronautics, telecommunications, electronics, scientific instruments) in which new products are the result of extensive research and development programmes. An important feature of the relationship beween science and technology is the new and increasing requirement for the management of the science-production cycle. Nevertheless, the administration of scientific activities needs a different approach from the well tried techniques of management. First, there is an important gap in the knowledge of the process of creating technologies. The stages of invention, innovation and diffusion not only give a logical order for technological knowledge, but also manifest quite different socio-economic environments and actors. Second, the more developed relationships between science and production are military, where the planning activities connect them into an integrated cycle. That is why the contemporary development of the science-production cycle is largely based on strategic military needs. Most of the advanced technological enterprises engage in projects financed by military, space or related agencies.

Competition and internationalization of capital

The development of productive forces is inherent in the accumulation of capital, creating quantitative and qualitative changes in the labour process of production by the reduction of labour value in order to get more surplus value: that is, a process of valorization of capital. In this sense, the technological change is 'capital biased' because it requires more means of production than labour. Despite this bias, identified by neo-classical economists as 'capital intensive', capital must circulate and rotate faster to counterbalance the tendency of falling rates of profit emerging from this increased use of capital-intensive technologies. As a result, activities connected with the circulation of capital, such as commercial banking and credit, are overdimensioned and biased, consuming part of the innovative efforts (marketing, publicity, over- diversification of products). But there are limits to the expansion of services involved with industrial productivity which point to the decrease of service productivity by the diffusion of

Long waves and international diffusion 201

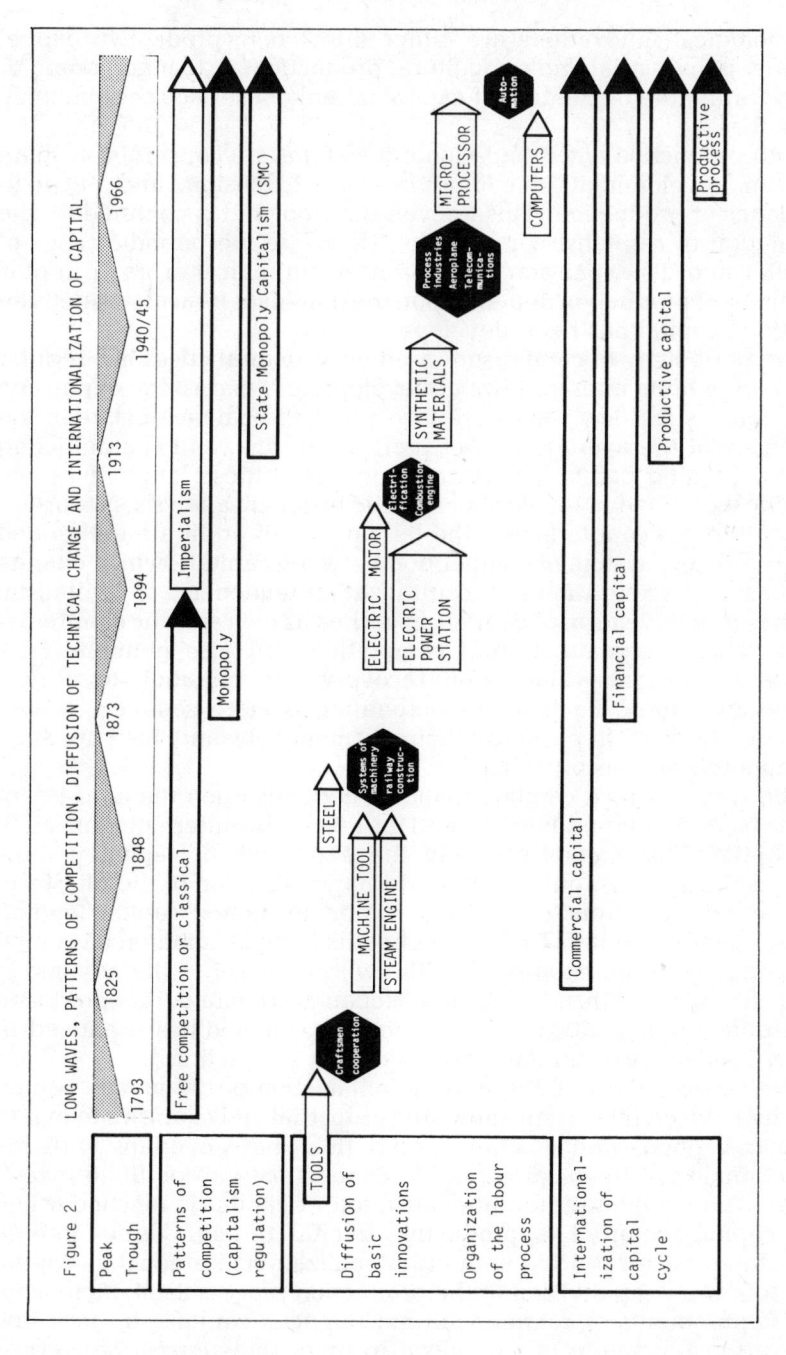

Figure 2   LONG WAVES, PATTERNS OF COMPETITION, DIFFUSION OF TECHNICAL CHANGE AND INTERNATIONALIZATION OF CAPITAL

technological innovations are either directed to productive capital (mainly in industrial and agricultural production) or to maintaining or accelerating the circulation of capital (mainly in service or commerce activities).

Both productive and circulation innovations are concerned with an uneven development of capital in time, extension, and between branches of production. This uneven development is confined by the regulation of capitalist competition. Therefore the periodic crises of capitalism or its expansion in different geographical areas or in new branches of production depend upon the dominant form of competition in which capital has been developed.

The periodicity of capitalism based on structural crises is a regular expression of its internal laws of development, that is, the expansion and depression long waves are a result of the balance between the tendency of the average rate of profit to decline, and counteracting forces. Turning points (peaks and troughs) of these long waves can only be explained in the light of specific historical analysis. Nevertheless the forces counteracting the falling rate of profit are embedded in a regulatory system of competition between capitals, which change under the concentration and centralization tendencies of capitalism. Each competitive form of capitalism carries its own specific counteractions. These regulate it until completion, but also generate new conditions for the valorization recovery and mutation to a new competitive form. Each time, competition is more complex, violent and fierce, from free competition, to monopoly and then to State Monopoly Capitalism (SMC).

The way in which capital circulates depends upon the adjustment made by each competitive form and the level of the internationalization of capital. This means not only the expansion of capital in new geographical areas, but also the internationalization of the phases of the capital cycle, from 'commercial' capital to 'money' capital, leading to 'productive' capital. This internationalization of commercial capital corresponds to the expansion of the world market to the advantage of the first industrialized countries, exchanging commodities produced in higher productivity conditions against commodities produced in lower productivity and often non-capitalist production.

The second phase is the internationalization of the money capital in which the credit system allows the industrialized countries to export industrial goods and machinery. But this phase overlaps with the development of financial capital in connection with portfolio investment. The complete internationalization of capital is reached when the capital exported is productive capital through direct foreign investments. This process of internationalization of productive capital will follow a deep division of the production process itself, increasing the requirements of commercial and financial capitals. In brief, the uneven development of capitalism in time, showing periodic crises related to the uneven extension of capitalism through the internationalization of the capital cycle, is based on uneven diffusion of the technology changes in the labour processes organization; the whole embedded in a specific form of regulation that mutates itself into new forms of capital competition.

Periodicity of capitalism based on its crises (historic-logic approach)

The economic history of capitalism has shown the appearance of periods of growth followed by a crisis, defining long waves of about forty to fifty years' duration. Apart from historical discussion in dating them, it is important to identify the turning points of the long waves.(5) Mandel's turning points are used for dating the sequence of upswing and downswing between the peaks and troughs of the development of capitalism.(6)

Cooperation between craftsmen was the technological basis for the first expansion of capitalism, dated from 1793 to 1825. The world market was then concerned mainly with mining, with some agricultural production from the colonies, the prevailing free competition being the main form of regulation.

Classical competition as such develops when the real appropriation of the labour process is reached through the use of the machine-tool. Nevertheless, conditions at that time were right for the introduction of the steam engine. Thus, the diffusion of machinery, or modern industry, was the technical base for the upswing during the period from 1847 to 1873. The steam engine as used in the locomotive and steamship (thus developing the transport infrastructure and encouraging the railways boom) facilitated the circulation of commercial capital.

Thus, productive forces expanded industrialization, aided by the development of new sources of energy. That is why the availability of electrical energy, the industrial production of electrical motors, and the introduction of the internal combustion engine to transportation are the basic technologies spreading in the capital expansion period from 1984 to 1913. The international flow of money increased substantially with the export of capital mainly for the exploitation of basic agricultural goods, mining and raw materials in the non-industrialized countries. Growing imperialism and the generalization of monopolies removed substantial profits from the colonies and semi-colonies to the metropolitan countries. In the post-war period from 1945 to 1966, the internationalization of capital was completed, beginning with the export of productive capital through forcing direct investments for local markets, both in industrialized and non-industrialized countries.

The export of machinery for this expanding industrialization and the increasing world market for raw materials are both operated by the multinational corporations. These giant enterprises, rooted in the more industrialized countries, get important support from their own states in looking after the general conditions of production. The 1945-66 expansion is very much linked to Keynesian policies, creating the capital 'environment' for its reproduction and accumulation. The economic functions of the state are mainly to ensure the reproduction of labour power, to participate in the managing of the money system, to redistribute the social surplus, and to guarantee the valorization of capital.

State Monopoly Capitalism assures the general condition of production by:

* regulating branches of production (such as electricity, water supply, transport and communication, infrastructure), services that socialize the indirect cost of labour (education, health, unemployment) and services for capital circulation (monetary system and commercial facilities);
* covering increased research and development costs and assuring the production of science and technology;
* offering opportunities for new profitable investments, nationally and internationally (armaments, environmental industries, infrastructure activities);
* dealing with international relations for defending and protecting the interests of international monopoly capitals.

In short, the state is playing a more important role as part of the counteractions to the falling rate of profit in a monopoly capitalism form of competition.

Crossroad for the next capitalist recovery

Once the expansion periods of capitalism have been roughly characterized, it is necessary to explain why the upswing turns into a depression (a trough) and the downswing turns into a recovery (a peak). The peaks are the end result of the counteractions to the tendency of the rate of profit to fall. The troughs in the long waves have been explained: by the clustering of basic innovations[7] (Mensch[8]); in relation to innovation life cycles and infrastructural investment fluctuations (Duijn[9]); as multi-factor causation (Freeman et al.[10], Schumpeter[11]); as the result of extra-economic causation factors (Mandel [12]); or as intrinsic to the capitalist economy (Kondratiev[13]). In these theories there is a first choice between intrinsic and exogenous causation and a second of the role of technical change in the upswing of capitalist production. There are two aspects to the upswing, internal and external, and what is important is the way in which capitalism provokes the conditions for the emergence of extraeconomic events (wars, revolutions, discoveries, innovations), and the way in which these events are internalized into the system's behaviour. Furthermore, a crisis develops its own countervailing economic forces arising from the devalorization of capital, as new possibilities for centralization of capital emerge (through the acquisition of poor profit-making enterprises, or bankruptcies) and from the concentration of products by investments in new 'capital biased' technologies. But the countervailing conditions can also arise from other extra-economic events, i.e. world wars, as in the 1913-1945 downswing. So the diffusion of basic innovations can induce a strong devalorization of old equipment reinforcing the diffusion of the new technologies. Summing up, the process of valorization and devalorization of capital caused by contradictory linked inherent and extra-economic variables is an expression of the uneven accumulation of capital in periods of expansion followed by periods of contraction of the economic activity.

In the present crisis, considering the previous internationalization of the capital cycle, the widespread knowledge of information technologies, biotechnologies and new materials, and the established SMC, the question is what recovery elements are there emerging from the crisis, and what are the minimal conditions and possible tendencies for the next recovery? Three main trends are considered: the internationalization of SMC, automation in social production and the new character of science as a direct force in production.

1. The internationalization of state monopoly capitalism as the emerging competition regulator

Crises develop both in international cooperation and conflicts. Conflicts, because the over-accumulation of capital is reducing not only new investments but the level of industrial capacity, and outlets have to fight for international markets. Cooperation, as when a compromise is reached between some states for a regulation of their international markets and investments (as is the case of steel in the EEC). The negotiations between states reflect competition between specific monopolies in which the state may or may not participate. That is why the first negotiations are often between firms, regardless of international regional state regulations.(14)

At the same time, international state negotiations are confronted with an internal conflict from those national production-oriented capitals looking for home protectionism against the international monopolies which derive benefits from the international agreements. Therefore the internationalization of SMC must be an external contradictory unfolding of the internal functions of the state. States are dealing with co-operative agreements on research, mainly for large projects in new energy sources (nuclear, solar), space communications, new materials, and military research and development projects. Whether inventions and innovations financed by the state, nationally or internationally, could be applied on a profit basis regardless of their social or military origin, is an open question, involving the patent system and new technological agreements.(15)

By far the most important aspect of state internationalization is its regulation of finance functions. First, the state regulates the rate of exchange with other currencies in connection with the balance between protectionism and openness of the economy. Second, it must keep a certain amount of control on the inflows and outflows of financial capital. Therefore, international agreements are part of a basic process of finding an international equivalent of labour value.

One possible scenario for the internationalization of SMC is a new regional basis for capitalism, not necessarily reflecting existing political and economic regional agreements (EEC, OECD, GATT, 77 countries, ALAC, ASEAN) but utilizing bilateral specific agreements. (These economic regions represent a new social basis for monopoly competition, through geographical, economic and political areas of influence.)

Despite the contradiction between socialism and capitalism, a new wave of expansion of production will encompass increased trade with the COMECON countries and a flow of investment agreements on a bilateral basis (Fiat in the Soviet Union, Japanese and US investment in China). The role of the semi-industrialized countries of the Third World is considered in the next section.

2. Automation in services and industrial production and a new international division of labour

The diffusion of automation would provide the technological basis for a future upswing of capitalism. Automation would have its first impact on services, increasing their productivity and indirectly liberating capital which can be used for the expansion of industrial production. Targets of office automation include government services, which are traditionally a large source of employment.

The pattern of diffusion of the automated labour process is uneven. For example, in the automobile industry there is a high degree of robotization in the press and welding lines, but not in the assembly section. In the electronic industry, too, assembly activies use relatively more labour. The division of labour in some productive processes locates the labour-intensive parts in countries like South Korea, Taiwan, Hong Kong, Mexico, Portugal, Brazil, producing backward industries. This means an internationalization of productive capital, where products are components of a larger process of production, and therefore not full commodities, but simply quasi-commodities priced by the interests of the international corporations.

The speed of new waves of mechanization and automation in traditionally labour-intensive activities poses the question of a reversal of this process in the more industrialized countries. This aspect and the automation of production will be considered in the next section.

The main barriers to automation appear to be the changes in the very basis of the capitalist society: (a) how can it continue the diffusion of automation and process industries (e.g. chemistry) in which the relation between constant capital to labour is very high, increasing the gap with labour-intensive activities; and (b) how can it continue the increase in production capacity through automation and robotization, thus increasing technical unemployment and the problems of 'realization' of production. So future scenarios will focus on changing the calculation of working-time value to a more flexible basis at the social level, connected with international deals between financial institutions.

3. The development of the science production cycle

Where can new jobs be created to support a widespread diffusion of automation in services and industry? If a possible upswing of capitalism emerges, then some jobs will come from classical compensation generated in the new branches of production related to automation and from the general expansion of capital accumulation, even with the expected increasing productivity of labour associated with new investments. Despite this, large employment opportunities, quantitatively and qualitatively, would be rooted in the quaternary activities: science, education, culture and art, and personnel services. Of all these, science, and the design and planning activities of social production are the most important. Although the generation of a scientific labour process of production is not imminent, space and military research activities show it to be planned in connection with their technical application. The creation of a science-production cycle, meaning that science could be planning for production purposes, is still only a possible future scenario.

However, work and employment issues will be a central aspect of the negotiations and contradictions between capital and labour involved in any capitalist recovery.

Role of semi-industrialized countries in a capitalist recovery

Overaccumulation of capital, appearing as an overproduction of commodities, is a problem of valorization of capital, because many capitals cannot achieve their profits. Crises of deficiency of demand are likely to appear, leaving unsatisfied societal needs both in industrialized and third world countries. In the latter the basic needs of food, housing, health, education and transport infrastructure are enormous, as well as in many less developed regions of the Western countries. For capitalist logic, the question is how some of these needs can be converted to effective demand, and also, how important this demand could be for capitalist recovery. Bringing effective demand to these people is a complex question, which cannot be regarded as a simple primitive accumulation in the sense of 'stages of development'. Nowadays, mankind has acquired the technical knowledge to increase land productivity up to a level to satisfy the food needs of many of the third world countries, and the industrial capacity to produce tractors and agricultural machinery is not fully utilized. But capitalism has laws that must be fulfilled in order to spread its previously developed forces of production. A basic assumption about unequal industrialization can be made: semi-industrialized countries are in a better position to be included in a new expansion of capitalism.

Identification of SICs is not an easy task because the concept itself is diffuse, indicating those countries that have built up a certain industrial capacity. This capacity is often oriented to consumer goods, to some intermediate raw materials, or to the extraction of natural resources. A general indication from selected groups of SICs suggests that a modest target of doubling the group average income per capita for 1990 (from $1,000 to $2,000 per capita) will correspond to 30% more than the actual GDP of the EEC countries ($3,688 billion for the SICs and $2,800 billion for the EEC). With an increase of one and a half times the 1990s per capita income for the year 2000 ($3000), the GDP of these SICs will correspond to the actual GDP of all the OECD countries (see Table 1). Besides the additional economic space that the semi-industrialized countries could represent in a possible upswing of capitalism, there are political and institutional trends that must be developed. One is the semi-industrialized SMC that has its own features and generally deeper contradictions than those in the industrialized countries. Another is the diffusion of new technologies provoking a larger differentiation in the production sector, destroying and conserving traditional activities. Yet another is the crisis in the international financial system, showing its limitations in the heavy debts of some semi-industrialized countries.

1. State monopoly capitalism in SICs

The main state functions in the accumulation of capital are biased by the dependency framed in a particular area of economic influence in which some multi-national corporations are operating. As can be seen in Table 2, in Latin America, 45% of the markets are supplied by the US. For South East Asia, Japan is the main exporter, providing 50% of their imports. In Africa, fifteen countries are getting half of their industrial imports from France. For India and the Middle East the markets are more divided. It is likely that in some of these areas the dominant influence will increase. This could be the case for US trade foreign policy, trying to link a closed North American 'region' by a

Table 1

PIB and population for group of selected semi-industrialized countries

	Countries (number)	Population (millions)	Income per capita ($000)	GDP (billions dollars)
Latin America	5[1]	269	2000	556
South-East Asia	8[2]	326	900	297
Asia	4[3]	733	600	436
Africa	4[4]	172	1300	224
TOTAL (1981)	21	1500	1000	1513
(1970)		1844[5]	2000[7]	3688[8]
(2000)		2140[6]	3000[7]	6420[8]
EEC (1980)[9]		270	10370	2800
OECD (1980)[9]		780	9675	7547

[1]Mexico, Argentina, Brazil, Venezuela, Colombia
[2]Hong Kong, Malaysia, Philippines, Singapore, Taiwan, Thailand, South Korea, Indonesia
[3]Saudi Arabia, India, Iran
[4]Algeria, Egypt, South Africa, Nigeria
[5]Estimated considering UNTAD estimations (2.0% per year), UNCTAD (1978) "Restructuring of World Industry", United Nations, New York.
[6]Ibid. (1.5% per year)
[7]Proposed
[8]Estimated
[9]OECD (1982) "Historical Statistics", Paris.

Table 2

TWC markets and the dominant exporter - 1980

Market		Dominant Exporter	% of the 6* big exporters to the TWC
Latin America:	Mexico	US	68
	Venezuela	US	47
	Argentina	US	31
	Brazil	US	34
South-East Asia:	Hong Kong	Japan	42
	Taiwan	Japan	57
	Singapore	Japan	37
	Korea	Japan	60
	Indonesia	Japan	54
	Malaysia	Japan	40
Africa:	Algeria	France	31
	Libya	Italy	29
	Morocco	France	50
	Kenya	UK	33

*US, Japan, RFA, France, UK, Italy.

Source: Le commerce avec le tiers monde: Les inconvenients des marches traditionels privilegies, "Problems Economiques", 14.4.82.

Table 3

Borrowing of TWC in short and middle term (billions/dollars)

	Total	Middle	Short
1978	59.3	35.4	6.5
1979	54.8	28.4	19.5
1980	49.0	19.5	26.5
1981	48.2	24.3	20.8

Source: "The Financial Times", 26 October 1982.

general economic agreement with Canada and Mexico. Thus the origin of a new wave of internationalization of capital in the SICs will depend on historic and new economic compromise with the industralized countries. That is why the possible economic links between third world countries depend on their role in the regions of influence in which they are located.

State monopoly capitalism in the SICs is, relative to their size, stronger than in the industrialized countries. This is because (a) the required level of capital for some activities can be reached only by the state or by important state participation; and (b) state participation in production allows it to offer the required infrastructure (transport, electricity, water supply), managing its social costs in order to provide an attractive 'profit climate' to national and international capital. In other words, the state can devalue a large part of the social capital in order to institute the valorization of capital, through a range of fiscal and economic policies (low energy prices, control of wages, forced exchange, subsidizing industry, monetary flow).

A permanent transfer of surplus to sustain a pattern of industrialization based on the subsidization of economic activity (imports, energy etc.) is fundamentally related to foreign debt. Each time the possibility of increasing the debt lessens and the cost of that debt increases in relation to exports (the service debt), the debtor turns to the short-term private banking system in order to deal with its service debt (see Table 3). Once the financial limit is reached the crisis is severe and the alternative for a new pattern of growth is dictated by the 'liberal' norms of the International Monetary Fund.

2. Diffusion of automation in SICs

Diffusion of automation techniques is related to the internationalization of capital and the internal economic and financial policies of the semi-industrialized countries. In the case of the internationalization of the process of production, some of the labour-intensive industries have been located in SICs (South East Asia and Latin America). This trend could continue, considering the strength of liberalization of national barriers for trade and money flows inside each region of economic influence. But if the trend towards automation continues, will the labour-intensive activities return to the industrialized countries? The answer depends on whether the SIC in question also provides economic and social conditions for the capital valorization of the automation processes. A change in the pattern of locating labour-intensive activities in the SICs will depend increasingly on technical integration of production introduced by automation. So the integration of production could take place either in the industrialized or semi-industrialized countries, depending on the global advantages of the multinational companies.

Diffusion of automation in SICs is not blocked by the neo-classical 'scarcity' of factors, so the industrialization of many sectors in the SICs has taken place on the basis of capital-intensive technologies (steel, petrochemicals, automobiles, textiles, food) where international and state firms are operating on an oligopolistic basis. Also, the so-called modern services are becoming 'mechanized' by the diffusion of computing services.

The result is a large heterogeneous level of technologies, from the craftsman to the very highly automated labour process organization. In this mass of technologies, the traditional ones play a role in the valorization of the more concentrated and technology-intensive capitals. Therefore, new automated industries are likely to be located in SICs, as a policy of attracting international investment and also because there is no resistance from the workers; either there is no labour organization for the new industrial activities, or there is stronger repression of the workers in these countries.

3. The financial system for SICs

As mentioned earlier, SICs are the weak link in the international financial system. This is not only because a bank could collapse, being unable to handle large credits to the larger debts of the country, but mainly because of the reduction of credit to those countries.(16) The substantial problem behind this contradiction is the international circulation of labour value through the SICs that has been until now focussed on the advantages of a club of industrialized countries. Therefore a possible scenario could be either a new international financial system with a higher participation of SICs, or some regionalization of the financial system with larger circulation of the currencies of semi-industrialized countries.

Notes

1. For a wide discussion on the logic-historical development of the labour process, see L. Corona, Revoluciones del proceso de trabajo en el modo de produccion capitalista, 'Investigacion Economica', 145, Facultad de Economica, UNAM, Julio-Septiembre 1978, pp.9-40.
2. 'The tool or working machine is that part of the machinery with which the industrial revolution of the 18th century started ... The steam-engine itself ... did not give rise to any industrial revolution. It was, on the contrary, the invention of machines that made a revolution in the form of steam-engines necessary' K. Marx, 'Capital', vol.I, reprinted USSR, 1977.
3. For Mensch, 'technical basic innovations produce new market and industrial branches'. G. Mensch, 'Stalemate in Technology. Innovations Overcome the Depression', Ballinger, 1979, p.47.
4. The products with a stronger technology density are usually defined as those with higher research and developments added value than the national average. See Conjoncture, 'Les echanges technologiques mondiaux', Paribas no.9, Octobre 1982.
5. 'The essence of every cyclical fluctuation model is the explanation of the turning

Long waves and international diffusion

points: why does an expansion turn into a contraction ' Van Duijn, 'The Long Waves in Economic Life', London, George Allen & Unwin, 1983.

6. The periods of the expansion and contraction of the economy give the following peaks: 1825, 1873, 1913, 1966; and the following troughs: 1793, 1848, 1894, 1940/45. See E. Mandel, 'Late Capitalism', Lowe & Brydone, 1975, pp.130,131.

7. The selected innovations are:
   Steam locomotive, 1830 (Liverpool & Manchester) GB
   Steamship (Atlantic crossing), 1838 (Sirus) GB
   Steel (puddling process), 1849 (Lohrge & Bremme) Germany
   Internal combustion engine, 1860 (Lenoir) France
   Electric power station, 1884 (Hollerith) USA
   Synthetic fertilizer (nitrogen), 1913 (BASF) Germany
   Cellophane, 1917 (La Cellophane) France
   Acetate rayon, 1920 (British Celanese) GB
   Polystyrene, 1930 (I.G. Farben) Germany
   Synthetic detergents, 1930 (I.G. Farben) Germany
   Synthetic rubber, 1932 (DuPont) USA
   Electronic computer, 1951 (Remington Rand) USA
   Microprocessor, 1971 (Intel) USA
   See Van Duijn, op.cit., pp.176-179.

8. 'The cycle begins with a technological stalemate resulting from stagnation in the formerly most highly developed industrial areas ... This situation engenders the cultural, political, social, economic and technological conditions required for the emergence of a cluster of basic innovations' Mensch, op.cit., p.68.

9. 'A complete explanation of the long wave therefore has to rely on the interplay of innovation life cycles and infrastructural investment' Van Duijn, op.cit., p.139.

10. 'We do not however subscribe to a theory of single-factor causation, nor to a conception of regular repetition with fixed periodicity of an unchanging cyclical mode of development' C. Freeman, J. Clark and L. Soete, 'Unemployment and Technical Innovation', Frances Pinter, 1982.

11. 'Now if we do ask this question (of causation) about all fluctuations, crises, booms, depressions ... there is no single cause' J.A. Schumpeter in R. Fels (ed.), 'Business Cycles: a theoretical, historical and statistical analysis of the capitalist process', abridged edition, New York, McGraw-Hill, 1964.

12. 'In the explanations of the sudden upsurges in the average rate of profit after the great turning points of 1848, 1893 and 1940/48, extra economic factors play key roles' E. Mandel, 'Long Waves of Capitalist Development', Cambridge University Press, 1980, p.20.

13. 'In asserting the existence of long waves and in denying that they arise out of random causes, we are also of the opinion that the long waves arise out of causes which are inherent in the essence of the capitalist economy' N. Kondratiev, 'The Long Waves in Economic Life', 1935, reprinted by Lloyds Bank Review.

14. 'In the past EC has attempted to promote industrial cooperation, but many large European-based companies have preferred making cooperative arrangements with Japanese and US companies rather than working with European competitors, notably in the field of automobiles and electronics' Mr Thorn, president of EC, Herald Tribune, 4, XII, 1982, p.1.

15. See the article 'Experts fear banks may grow too cautious', International Herald Tribune, 25, 26 September 1982.

12 Long cycles and the international diffusion of technology

Luc L.G. Soete

The dramatic resurgence of interest in long waves has not been confined to the profession of the economic historian, but seems, in a typical Schumpeterian fashion, to have led to a bandwagon effect, involving not only economists and social scientists of all disciplines, but also businessmen, technologists, scientists, stockbrokers, and even patent agents. Not that this seems to have lent any more credibility to long wave theories. Long waves and even 'long term movements' (implying no specific cyclical periodicity, but a process of long-term accelerations and decelerations in growth with the periodic occurrence of crises) remain something of an aberration to the academic profession(1) and to policymakers(2) in general.

Yet the study of and evidence for long term movements is a relatively well-established research field within economic history analysis. Economic historians generally distinguish between the pre-industrial era (pre-1790) and the industrial era (post-1780). The latter is primarily related to technological changes and accelerations in the production of industrial goods,(3) whereas in the former, short-term and long-term movements are in the first instance related to meteorological and climatological factors affecting agricultural production. As pointed out by Mauro,(4) the fundamental difference between the two lies in the fact that in the pre-industrial era one is not confronted with overproduction of industrial goods, but underproduction of agricultural goods. Interestingly, the evidence in relation to these pre-industrial long-term cycles seems rather conclusive and the relationship with meteorological factors significant.(5) The time span of these pre-industrial long-term cycles (approximately 30 years) appears, however, to be totally different from the industrial long waves.

Most of the current debate on Kondratiev long waves still centres on the two fundamental questions raised by Kuznets some 40 years ago in relation to the industrial long-wave phenomenon, i.e. (a) what empirical evidence is there in support of such cycles, and (b) what are the possible explanations for such a cyclical pattern. In contrast to the pre-industrial debate there has been over the last 50 years relatively little progress on both these issues.

(a) On the question of evidence, it is somewhat paradoxical that with the elaboration of more sophisticated statistical methods, such as spectral analysis, there is not only more controversy about the existence of industrial long waves, but there is actually no longer agreement about the nature of the present phase, with Rostow(6) arguing in terms of a Kondratiev upswing and most other Kondratiev believers arguing in terms of a Kondratiev downswing.(7)

The statistical problem is in many ways unsolvable and relates to the fact that empirical evidence over the last 200 years, apart from its scarcity and unreliability, is of very little use in allowing one to test in a statistically significant way the existence of 50 year cycles.

(b) Regarding explanations for long waves, a large variety of theories ranging from purely monetary and cost price explanations to Schumpeterian and Marxist interpretations, have been competing with each other in trying to identify the major causes of the cyclical upswings and downswings. This luxury in theoretical explanations is in sharp contrast to the poverty of the industrial long wave evidence. Consequently, the question as to which theory provides the 'best' explanation remains at present of little relevance.

From these comments, it will have become obvious that there is not much point in pursuing these two questions in the traditional mechanistic statistical way in which they were initially posed. A more or less structural approach seems to impose itself, focusing on the role of some of the variables closely related to each of the theories mentioned above. These explanations of the causes of the major booms and depressions which have struck the Western world over the last 200 years are free from any of the fixed periodicity assumptions implicit in long wave theories. This is not to deny though the value of much long wave research.

Even if the long wave remains largely unproven, the multitude of theoretical views in relation to the cyclical nature of the 'myth' or the 'phenomenon' remains an area rich in speculation and insights.(8) Industrial long cycles remain a question of belief, largely because of the present statistical impossibility of proof. Its 'belief appeal' relates no doubt to the remarkable fact that the theory was first put forward in an upswing phase some 60 years ago and that most of the debate still centres around the evidence of that time, i.e. the evidence for the 18th and 19th centuries. If anything, the past 60 years have proved the strongest support yet for Kondratiev long-term cycles.

Within its broad economic framework, the debate on long waves has centred around the question of whether the evidence relates primarily to the real sector or is just a monetary or price phenomenon.(9) Because of the abandonment of the gold standard in the 1930s, emphasis on price movements as a result of gold production and gold reserves has lost much of its relevance for present 20th century long-term cycles. Furthermore, the strong inflationary pressures over the post-war period, contrasting sharply with the 19th century trend in prices, have led to a shift of focus to 'real' (production, investment, real wages, etc.) factors as being the most suitable indicators of long-term movements.

The focus in this paper being on the long-term economic development of the presently industrialized and semi-industrialized countries, long waves will in the first instance be reflected in 'real' factors. Furthermore, to the extent that it is technical change which has been the critical factor in the industrial development of these countries, our approach will largely be within the Schumpeterian innovation/long-wave tradition, as exemplified in much of Freeman's recent writings,(10) Such an approach fits well within many other long-wave contributions, such as Mandel.(11) It has, however, a number of distinct features in relation to the role of technology as the crucial factor behind (both explanatory of and dependent on) long-term movements of the economic system - not just technology or technological change in its most narrow form (the rate of advance of industrial knowledge), but primarily the broader concept of technological change, including the actual diffusion of existing technologies, whether originating from within the country or from abroad. One of the aims of this paper is indeed to arrive at a further integration of a number of technology diffusion theories, both in relation to inter-firm and inter-country technology diffusion and Schumpeterian long-wave theories. While the 'domestic' inter-firm diffusion of clusters of innovation, referred to as 'new technological systems' is most strongly emphasized and indeed the central concept in Freeman et al.'s long-wave theory (1982, chapter 4), the role of foreign technology and the international diffusion of technology remain largely outside their theoretical mode. Yet, as they emphasize in chapter 8, the importance of foreign technology and the international diffusion of technology is historically a well recognized factor in the industrialization of both Europe and the United States in the 19th century, and even more strikingly of Japan in the 20th century.

In this paper we wish to analyse how the process of the international diffusion of technology fits historically within the overall process of the various expansion waves which have characterized the advanced industrialized countries over the last 200 years, and what lessons there are to be learnt by presently-industrializing countries.

To do so, we first briefly summarize the argument in relation to 'world' long cycles or movements and how national long-term industrial development depends upon and influences such world cycles. In a second section we briefly review some of the most important contributions to the technology diffusion literature, particularly the further elaborations to the 'standard diffusion model' and their relevance to structural (particularly Schumpeterian) theories of industrial development and long waves. In a third section, we open the discussion to the international diffusion of technology. We conclude by emphasizing some of the implications for semi-industrialized or newly-industrializing countries and the possibilities for a more rapid world-wide diffusion of the new microelectronics paradigm.

World long waves or country-specific long waves?

There is probably broad agreement amongst most long-wave writers that long waves relate in the first instance to the capitalist (or socialist; see Pasinetti(12)) world economy. As Schumpeter himself (13) pointed out: 'Capitalism itself is, both in the economic and sociological sense, essentially one process, with the whole world as its stage.' Kuczynski(14) has dated the emergence of a world capitalist economy at around the mid-19th century: 'Since about 1850 there has existed a capitalist world economy in the true sense of the word, the world economy has come to predominate over national economies. This predominance first became evident during the world economic crises of 1857.' The precise timing of the national predominance of a world capitalist economy is largely an academic question. As emphasized by many economic historians and development economists, particularly Wallerstein (15) and Frank(16), in pre-industrial commercial capitalism, trade and financial exchanges between Europe and its colonies were sufficiently significant to warrant the description of a world capitalist economy.

The increase in world-wide international trade, investment and financial flows is, however, a distinct feature of the industrial era. In contrast to the pre-industrial era, these international flows, particularly those emerging from the leading technological countries were generally based on a significant technological advantage - an advantage which would take the form of more cheaply produced goods or new commodities. While such trade flows had significant disruptive effects on many of the importing countries, the overall long-term effect resulted in a more interwoven world (capitalist) economy where each country's long-term cycle became increasingly dependent on the world economy's long-term movement.

The dramatic disruption in trade in many European countries, related to Britain's early technological trade advantage over the first and second Kondratievs, led to a quite distinct international feature of long waves: that of the crucial importance of the import of foreign technology and the international diffusion of technology for autonomous 'national' growth and technological leapfrogging. The crucial point about this feature, as pointed out by Freeman et al. (chapter 9), is that it basically is unrelated to the overall long-wave upswings and downswings. It is primarily the result of an internal, country-specific, autonomous growth process nourished by the existence of a technological gap with the world technological frontier, where the technological gap can be thought of as a 'pump of diffusion' to use Gomulka's(17) terminology. The implications for the long-wave debate of this international technology diffusion feature are important. Freeman et al., for example, point out that it is primarily this autonomous growth feature which obscures much of the long-wave evidence, based as it generally is on individual country evidence. A similar point is made by van Duijn(18):

> National economies have their own 'life cycle of development'. Depending on their take-off date, countries may perform strongly during depression or they may do rather poorly during a long-wave expansion. In the world economy as a whole, the effects of extra-rapid and extra-slow growth cancel each other out, but the fundamental causes of long waves remain. Over the years, basic innovation life cycles become international phenomena.

To summarize: over time, long waves become increasingly an international phenomenon. International trade and other financial flows reinforce the international dependence of individual countries whether capitalist or socialist, rich or poor, on 'world economy' long-term (and short-term) movements. Catching-up and technological leapfrogging based on the international diffusion of technology, by contrast, is largely independent of world cycles (short or long).
Before turning to a more detailed discussion of the interaction between international technology diffusion and international long waves, we review briefly a number of micro-economy theories of technology diffusion and their relationship with industrial growth.

Inter-firm technology diffusion and industrial growth

The traditional technology diffusion model

Theories of inter-firm technology diffusion rely heavily on the mathematical theory of epidemics to explain the so-called 'retardation hypothesis' The fundamental reason why individuals or firms do not immediately adopt a new innovation relates both to uncertainty and lack of information about the new technology. While 'potential' adopters know about the innovation, they probably know too little about its actual performance within their own economic environment. Observation of the experiences of prior adopters will, however, reduce their technical and market uncertainty. In other words, it is learning from the experience of others which leads to imitative behaviour and the bandwagon effect and which provides the underpinning of the logistic S-shaped diffusion curve, identical to that of the spread of diseases. This curve predicts that the proportion of individuals or firms having adopted the innovation will increase at an accelerating rate until 50% adoption is attained, whereafter adoption increases at a decelerating rate and total adoption is approached asymptotically.

While the model offers substantial advantages in allowing one to relate the speed of diffusion to the economic characteristics of the innovation and its adoption, in particular its relative profitability and the required investment, a large number of criticisms can obviously be put forward in relation to its simplicity and 'mechanical' features.

The S-shaped diffusion pattern is in many ways rather similar to the industrial growth pattern of industries depicted by Schumpeter in both 'The Theory of Economic Development' (1912) and 'Business Cycles' (1939). This is not surprising; the concepts of 'imitation' and 'bandwagons' so crucial to the diffusion literature, are indeed central concepts too in Schumpeter's theory of the long-term rise and fall of industries. The formal link between the two sets of theories is made in Freeman et al. (1982, ch. 4). In the latter analysis it is precisely the notions of clusters of innovations, so-called 'new technology systems' including the follow-up innovations made during the diffusion period, which are linked to the rapid growth of new industries and might provide the ingredients of an upswing in overall macroeconomic growth. In our diffusion terminology used here, this can be viewed as an 'envelope' of diffusion curves from a set of closely interrelated clusters of innovations. Occurring within a limited time span, these might tilt the overall economy in the early diffusion phases to a higher level of economic growth, and provide the underlying explanation for the overall long-term S-shaped development pattern of new industries and the overall long-wave development trend.

Another similarity with diffusion models can be found in Rostow's theory of the stages of economic growth,(19) again with a distinct S-shaped pattern of take-off, rapid growth with the 'drive to maturity' and the 'age of high mass-consumption' and standardization. Rostow phases contain many of the S-shaped development patterns assumed to exist for new products, as typified in the marketing and subsequent international trade literature on 'product life cycles'

Within the development literature, Rostow's theory has been criticized most strongly; the mechanistic, quasi-autonomous nature of the process of economic growth has even been branded as ahistorical on some occasions. Interestingly, though, the critique on the mechanistic nature of Rostow's growth model has, as indicated, been one of the major issues in much of the current long-wave debate, and is also a major concern in most of the recent diffusion literature, criticizing the simple, mechanistic 'epidemic' technology diffusion model. This literature actually provides a number of interesting insights into both the broader industrial growth theories and the more specific long-wave theories.

Recent elaborations on technology diffusion models

The first area of critique on the 'standard' diffusion model has led to the application of 'probit analysis' to the traditional diffusion model. This was already a well-established technique in the study of the diffusion of new products between individuals and can be extended in a relatively straightforward manner to inter-firm diffusion. The central assumption underlying the probit model is that an individual consumer (or firm) will be found to own the new product (or adopt the new innovation) at time T if his income (or size) exceeds some critical level. As Davies[20] says:

> This critical, or tolerance income (or size) represents the tastes of the consumer (the receptiveness of the firm) which in turn may be related to any number of personal or economic characteristics.

Over time, though, with the increase in incomes and assuming an unchanged income inequality, the critical income will fall with an across the board change in tastes in favour of the new durable, (due to imitation, greater and better information, bandwagon effects, etc.). The advantages of the probit or threshold level diffusion model relate both to the possibility of introducing behavioural assumptions concerning the individual consumer (or firm) and:[21]

> the reduced need for ad hoc theorizing on the determinants of the speed of diffusion ... diffusion will be more rapid the faster the growth in incomes, the faster the bandwagon effect, the more equal is the income distribution and the less variable are consumer tastes.

For firms, it is generally assumed that firm size is the most important factor behind the inter-firm differences in the profitability of adoption. Both the ability to acquire and understand the technical information needed to assess the new innovation and the readiness to accept risks are assumed to be related to firm size.

The probit model bears some relevance for industrial growth theories. A 'critical' per capita income level, for instance, is a concept which can be introduced in a straightforward manner in Rostow's stages of economic growth. Replacing the concept of individuals with that of countries, different behaviour between countries in their growth performance can be explained and expected. Considering both the extreme variation in a country's ability to take risk and assess new innovations (the variation in consumer tastes in the probit model), and the extreme levels of income inequality at world level, it should come as no surprise that industrialization at the world-wide level (diffusion) has been slow, and that many poor countries, even with the fall over time in the 'critical income' industrialization level, have never reached the stage of take-off.

In terms of the Schumpterian model of long-term structural change and long waves, the probit model also offers interesting insights into the slowness of diffusion once a cluster of innovations (the new technological system) has occurred. In the depressive phase of the long wave, few firms are ready to take the risks or are in a position to assess the profitability of adopting the radically new innovations. The critical yardstick or target against which the pay-off period associated with adopting the new innovation will be compared (as in the probit model (22)) will be high, and the overall slow growth in the economy will further retard the rapid diffusion of the new technological system.

The second major area of criticism of the standard diffusion model relates both to the static nature of the latter and the pure demand nature of the model. Metcalfe(23) in particular has emphasized the limits of the standard model in this area:

> There are well-documented reasons for expecting both innovation and environment to change as diffusion proceeds. Improvements to the innovation, general economic growth, changes in relative commodity and input prices, other complementary or competing innovations can all be expected to occur during diffusion ... Foremost among these endogenously determined changes will be induced developments in the technology and induced changes in the profitability of adopting the innovation.

> This suggests a crucial weakness in the standard diffusion model in that it focuses attention entirely on the demand for an innovation by potential adoptors. The supply side is ignored and with it the question of how the profitability of adopting and producing the innovation is determined. Profitability influences the pace of diffusion but equally the pace of diffusion will influence profitability.

In Metcalfe's diffusion model the price of the new innovation is no longer a constant, or evolving along a particular time path, but is determined itself by the process of diffusion. In addition, supply of the innovation is limited by productive capacity, the rate of increase of which depends on the profitability of producing the innovation. A typical Schumpeterian scenario of an entrepreneur innovator emerges, whose temporary reward consists of the initial monopoly profits, which are gradually competed away as imitation takes place and the innovative potential is exhausted. At the same time, though, as the rate of return of the innovator and imitators falls, 'the associated reductions in price increase the profitability of adopting the new innovation' which is further diffused.

There are various other illuminating aspects in Metcalfe's diffusion model: the role of secondary or incremental innovations which can both be autonomous and induced by the diffusion process, which will further expand the diffusion potential of the initial innovation; the importance of past investment in, and existing commitment to, the technology which is being displaced, slowing down the diffusion of the new innovation ('new technology competes on disadvantageous terms' ; and the importance of inter-technology competition between the existing and new technologies. Rosenberg in particular has emphasized the importance of improvements to existing technologies as a result of the emergence of a radical alternative technology. Thus it has often been alleged that the diffusion of steam power in the last century was retarded by a series of improvements to existing water power technology which further prolonged the economic life of the old technology. The process of a 'dying technology' is indeed a slow process, with the old technology firms often living off fully-recovered investment and sometimes being able to underprice the innovation-adopting firms (e.g. the present UK foundries industry).

Metcalfe's diffusion model offers, of course, extensive scope for integration in the Schumpeterian model of long waves, as analysed in detail in Freeman et al. (1982). Broadening the discussion from a discrete innovation with its specific innovation path, to the concept of a cluster of interrelated innovations and the emergence of new industries, it can be shown precisely how profitability for both suppliers and adopters is the crucial ingredient in the overall long-wave pattern of economic growth typical of the Schumpeterian model.

> In Schumpeter's model the profits realised by innovators are the decisive impulse to surges of growth, acting as a signal to the swarms of imitators. The fact that one or a few innovators have made exceptionally large profits does not mean, of course, that all the imitators will do so. It is enough that they hope to, or even that they hope to make a fraction of them. As the bandwagon begins to roll profits are competed away and some people fall off the wagon. Schumpeter himself stressed that changing profit expectations during the growth of an industry are a major determinant for the sigmoid pattern of growth. As new capacity is expanded at some point (varying with the product in question), growth will begin to slow down. Market saturation and the tendency for technical advance to approach limits, as well as the competitive effects of swarming and changing costs of inputs, may all tend to reduce the level of profitability and with it the attractions of further investment.

Other aspects of Metcalfe's diffusion model, in particular his emphasis on the initial disadvantageous terms on which the new technology competes and the importance of inter-technology competition, offer similarly interesting insights in the broader process of long-term structural change, economic growth and long waves. Thus, while the new innovation cluster might provide a radical improvement on existing technology, past investment outlays in the latter (both in terms of capital and skills and - one could add - in terms of R&D oriented towards incremental improvements to existing technologies) will provide a powerful retardation factor in the diffusion of the innovation both from the supply and demand points of view. This might well explain why in Schumpeter's earlier work ('The Theory of Economic Development' 1912) radical new innovations were identified with new firms, set up by 'exceptional entrepreneurs' rather than with existing large firms. In Freeman et al. (1982) it is argued that the early Schumpeter model will be particularly relevant in the trough and early long-term upswing of the long wave, closely related as it will be to the growth of new industries and technologies.

The concept of inter-technology competition, on the other hand, offers an interesting extension to Nelson and Winter's concept of 'national trajectories' and Dosi's[24] notion of technological paradigm. The emergence of a new technology has well-known implications for the 'natural trajectory' of the existing technology which might receive a new impulse (the so-called 'sailing ship effect' ; the new paradigm might actually be defeated. There are many cases of major new innovations which, after an initial success, have been 'defeated' by crucial improvements to existing technologies (e.g. the Wankel engine, electric cars, Concorde, etc.).

The discussion so far has been limited to the diffusion of 'domestic' technology within a closed economy. In the next section we turn to the issue of the international diffusion of technology and long-term economic growth.

International diffusion of technology and long-term industrial growth

Turning now to the issue of the international diffusion of technology, many of the points raised in the previous section can be extended in a relatively straightforward manner to the international scene, where countries replace the notions of innovating and adopting firms of the previous section.

The international diffusion of technology has been a major factor behind most industrial nations' economic growth. However, from the innovating, technologically leading country point of view, the major issue will be (just as in the case of the Schumpeterian entrepreneur-innovator) how to prevent the competing away of its international technology monopoly position. It can either try to maintain a continuous flow of innovations through heavy expenditure on research and development, or try to appropriate the new technology and limit its international diffusion. It is worth noting, for instance, how Britain in the early 19th century tried unsuccessfully to limit (through prohibiting the export of machinery) the diffusion abroad of the technology which formed the basis of its technological and economic lead over the first long wave.

In the long run, though, both policies will probably fail to guarantee continuous technological leadership. Imitators will appear (international 'appropriation' is limited), returns on the technology will fall, and radically new technologies, new 'technological paradigms' might emerge. While the latter will probably originate from within the technologically leading country, the internal national diffusion will be hampered by the various factors mentioned earlier, the new technology competing (in its diffusion) on disadvantageous terms. Thus the possible previous investment outlays in the existing technology,(25) the commitment to the latter from management, the skilled labour force and even the development research geared towards improving the existing technology, might well hamper the diffusion of the major new technology to such an extent that it will diffuse more quickly elsewhere, in a country uncommitted, both in terms of production and investment, to the old technological paradigm.

The industrialization in the 19th century of the United States, Germany, France and a number of smaller European countries provides ample support for this view.

The second half of the 19th century indeed saw Great Britain being overtaken as the most important growth and innovation pole. This was linked both to the successful imitation and catching up of a number of European countries and the United States (primarily through import substitution and protectionism based on 'infant industry' ideas), and to the decline of some of the major industry carriers of the industrial revolution (such as the cotton industry, in which the UK had retained its lead). In the most important growth industries, primarily steel and metalworking, Britain - while having initiated some of the most important innovations (the Bessemer converter, Siemens' open hearth in 1866, and the Thomas process in 1878) - lost its innovative lead in exploiting these to Germany, which overtook Britain in the production of iron, coal and steel by the end of the century.

The turn of the century sees the emergence of both the United States and Germany as major growth and innovation poles. The UK was clearly lagging behind with respect to the major innovations which emerged in this period (electricity and automobiles). With respect to electricity, it was primarily the United States (Edison's electric power steam central in 1882, Westinghouse's alternating current distribution of electric power in 1885) which took the lead; its production of electricity was about five times that of the UK by 1907, while its total industrial production was about the same.(26) With respect to motor cars, the lead was both European (Germany with Daimler in 1887 and Krebs in 1892, and France with Panhard and Levasseur in 1892-4) and American (Duryea in 1892). By 1910, according to Ray,(27):

> the largest car maker in Britain was Henry Ford, producing more cars than the next two largest firms combined. Thus, British entrepreneurs lagged in this major industry too, and this is likely to have had a considerable impact on the general development of a number of branches of the engineering industry in view of the requirements of automobile production in terms of machine tools and new machines - as well as some other newer and older industrial sectors such as rubber and instruments, or even textiles and timber.

This dramatic change in fortune from absolute technological leadership, producing for example more steam engines in the mid-19th century than the whole of the rest of the world put together, has of course many causes, but it is no doubt related to the rapid international diffusion of British technologies, which were hampered in their internal domestic diffusion.

From the technology-adopting point of view, the issue is not just one of how quickly one adopts. The profitability of adopting, apart from the factors emphasized in the standard diffusion model (the investment required, the profitability of the innovator) will also depend on the 'critical' income level (as in the probit model), the technology supply market, and the maturity of the technology which is being diffused.

To the extent that a continuous flow of improvement innovations probably will occur over the early diffusion phase, immediate adoption will not only imply a significant contribution to the monopoly profits of the supplying innovator, but also a commitment to a technology which might well undergo significant improvements in the not too distant future.

This, then, pinpoints some of the advantages of late industrializers, both in terms of catching up with present technological leaders and in terms of acquiring foreign technology at a more competitive price. This has been most obvious in the case of the rapid industrialization of Japan in the 1960s and 1970s, where world 'best practice' productivity levels were achieved over a very short time in steel, cars, electronics, numerically controlled machine tools, etc., largely on the basis of initially imported technology. Could it be that a similar pattern is emerging in a number of so-called 'newly industrializing countries'

Elsewhere[28] it was argued that with the fourth Kondratiev downturn, with more countries than ever having drawn nearer to the technological frontier following the exceptional period of convergence/technology gap growth over the postwar era, a period has developed of intense technological competition between the major contenders for technological leadership. This implies a continuing increase in autonomous research spending (as opposed to the preponderance of research spending auxilliary to technology-import in the early catching-up phase) in the 'catching-up' countries - yet more and more duplication of research. At the same time, the international diffusion of technology increases more rapidly, the various technological leaders competing against each other in domestic as well as foreign markets. The result is a further decline in the rate of return to inventive activity and innovation. Even the smaller industrially advanced countries, seeing their productivity growth more and more restrained by the rate of advance of the world technological frontier, will further increase their autonomous research effort, in order to achieve technological leadership in those sectors in which they have developed some comparative advantage in either technology or export.

Conclusions

There is, we would argue, broad agreement between long-wave believers and long-wave disbelievers that a period of recession or (even more so) depression is in the first instance a period of a dramatic 'structural' shake-up. A priori, there is no reason why such a shake-up should be a purely domestic phenomenon. With the downturn of the long wave there is also a structural shake-up in the ranking of countries, in terms of both growth and technological performance, similar to the structural shake-up in the internal and world-wide ranking of industries and firms.

As hinted at in the first section of this paper, such a shake-up is primarily a function of two distinct, sometimes contradictory, tendencies within each industrial country. On the one hand, the existing international trade and financial relations between the domestic economy and the world economy will be crucial in determining the severity of the domestic recession and depression; on the other hand, the autonomous internal industrialization or catching-up growth potential (based on the existence of a significant technological gap) and the international diffusion of technology, while being retarded, will remain largely intact and provide a strong counteractive influence to the world-wide downturn pressures on the country's growth performance.

The point can be illustrated with reference to the present growth performance of semi-industrialized or newly industrializing countries.

Those heavily committed to an export-led industrialization strategy in manufactured goods, and even more in recent years those dependent on oil-exports, have suffered significantly from the world-wide recession - their increased integration into the world economy leading to an increased dependence on world-wide cycles, yet their internal growth potential, while retarded and hampered by international financial constraints, remains largely intact. As a matter of fact, their growth performance is at present significantly higher than that of most industrially and technologically advanced countries, whose individual internal growth support measures, e.g. through inflation, are wiped out in no time by the overall world-wide recession.

This should clarify some of the contradictory conclusions which sometimes emerge out of the long-wave debate and its implications for the Third World. Indeed, depending on one's viewpoint, the emphasis can be on the influence of the world-wide recession on the newly industrializing countries' growth performance,[29] emphasizing the increasingly close interdependence, the phenomenon of increased protectionism in the advanced industrial countries, the possibility of trade reversals linked to technical change, the role of multi-national corporations, etc., or one can emphasize the relatively autonomous nature of the industrialization process in these countries, based on catching-up and technological leapfrogging.

The reason for a greater emphasis on the latter than on the former in most of our own work, and also in Freeman et al., relates precisely to the importance given to diffusion and the concept of a new technological system or new technological paradigm in the analysis of the long wave. Thus, whereas the specific diffusion retardation factors are in the first instance economic in nature, there will exist institutional country-specific retardation factors which might, in the case of a new technological paradigm, be as severe for the technological leader as for late industrializers.

The point can readily be illustrated with reference to the present new microelectronics paradigm, the rapid diffusion of which can be expected to lie behind a new long-term upturn. Time and space prevent discussion of the issue here in much detail. However, as has been emphasized in most of the microelectronic literature, the specific features of the new technological system, in particular its potential for small scale and its dramatic effect on capital productivity,[30] provide a set of strong incentives for its rapid diffusion not only in advanced, but even more in semi-industrialized or newly industrializing countries, where growth has been hampered by general capital shortage problems. Furthermore, the deskilling effects of the new electronics paradigm appear to be particularly severe in relation to a wide variety of highly specialized technical (mechanical and electrical) skills, which at this moment form the major specific human capital bottleneck in most less industrialized countries. As a consequence, one might expect there to be more social resistance vis-a-vis the new paradigm in the advanced countries.

Finally, and as emphasized by Teubal,(31) electronics technology differs fundamentally from technical change in both process industries (materials, chemicals) and mechanical and electrical industries, in so far as technical change in electronics is more directly related to the underlying scientific and technical/educational learning. In both process and mechanical production, technical change is in the first instance based on productive learning-by-doing, e.g. the setting up of proces plants or the designing of new machine tools. Again, to the extent that scientific and technical education is not as crucial a bottleneck in semi-industrialized or newly industrializing countries as productive learning and experience, the micro-electronics paradigm might well diffuse more rapidly in these countries than is generally expected.

All this, added to the increased world-wide technological competition mentioned earlier - from a predominantly seller's market to a more competitive and, in some specific technology areas, predominantly buyer's market (e.g. electric power generating equipment(32)) - and the specific difficulties in legally appropriating world-wide electronics knowledge (see, for instance, the debate about the protection of software, which at present is only covered through copyright protection), suggest that the international shake-up linked with the present downturn might well lead to a different set of growth and innovation poles than the ones we have been witnessing over the last Kondratievs.

This debate remains however extremely speculative. There is less speculation involved in the argument that the major world institutional factor preventing a more rapid upturn in world economic growth, based on the world-wide diffusion of some of the major new technological systems, relates to unequal world income distribution. This results in the present paradox of a lack of 'world demand' despite the fact that the great majority of the world's population is not even able to see its most basic needs fulfilled. In many ways, just as Keynesianism was in the thirties a (very late) institutional response to the regular over-production crises of the 19th and early 20th centuries, based on the lack of internal demand in the various European countries, there seems to be a clear need for some sort of institutional change - international Keynesianism - in response to the lack of world demand, and specifically related to those institutional financial factors which at present prevent the greatest part of the world's population from expressing its 'effective' demand (see in particular the Brandt Commission proposals).

How such a world institutional change can be brought about remains, however, a very open question, and it is to be feared that only a further deepening of the world depression will bring about sufficient pressure on the presently advanced countries to force them into such a radical institutional change.

Notes

1. As Van Duijn observes, long waves are dealt with in a one sentence footnote (p.241) in the latest edition of Samuelson's economic textbook. J.J. Van Duijn, 'The Long Wave in Economic Life' London, Allen & Unwin, 1983.
2. See S. Brittan's recent 'economic viewpoint' in the Financial Times on the 'Myth of the Kondratieff'
3. Where the specific short-term Juglar and long-term Kondratiev cycles coexist with the broader 'secular trends' - 200 year cycles - related to the 'life cycle' of an economic system or even in broader terms, civilization.
4. F. Mauro, 'Les mouvements longs dans l'histoire de l'amerique latine: problems poses' 1983.
5. J. Georgelin, 'Venise au siecle des lumieres (1669-1797)' Paris et la Haye, 1978; J.G. Grenier, 'L'Utilisation de l'analyse spectrale pour l'etude des series de prix dans la France pre-industrielle (XVIe-XVIIIe siecles)' 1983.
6. W.W. Rostow, 'The World Economy: History and Prospects' Austin, University of Texas Press and London, Macmillan, 1978.
7. For some clarification about this controversy see I. Wallerstein, Kondratieff Up or Kondratieff Down?, 'Review, vol.II, no.4, Spring 1979, pp.663-673.
8. See in particular the arguments about long waves in the development of philosophy (e.g. R. Klos, Philosophy and Long Waves - some summarizing, in 'Cyclical Fluctuations: an interdisciplinary approach' 8th Intl Economic History Congress, Budapest; Netherlands, Free University, 1982), strike patterns (e.g. P. Boomgaard, Strike patterns and economic fluctuations: the building trade in the Netherlands, in 'Cyclical Fluctuations' op.cit.), proletarian insurgencies (e.g. E. Screpanti, Long economic cycles and recurring proletarian insurgencies, 'Review' 1983), or even patent laws (e.g. W. Kingston, 'An Innovation Certificate?' 1982).
9. Only J. Delbeke ('Towards an endogenous macro-economic interpretation of the long wave, the case of Belgium, 1830-1980' paper presented at the 3rd Conference of the Council for European Studies, Cycles and Periods in Europe, Past and Present, held in Washington, DC, April 29 - May 1 1982; and 'The interdependence of real and monetary factors in a long wave perspective' paper presented for the colloquium 'Research on Long Waves' Paris, March 17-19 l983) has effectively tried to integrate both approaches and analysed (rather successfully) the interdependence between the real and monetary factors in his study of long waves in the Belgian economy.
10. C. Freeman, Innovation and the process of economic growth, and Innovation as an engine of economic growth: retrospect and prospects, in H. Giersch (ed.) 'Emerging Technologies: Consequences for Economic Growth, Structural Change and Employment' proceedings of Kiel Symposium 1981, Tubingen, Mohr, 1982, pp.1-32. C. Freeman, J. Clark and L. Soete, 'Unemployment and Technical Innovation: A study of long waves and economic development' London, Frances Pinter, 1982.
11. E. Mandel, 'Late Capitalism' Frankfurt, Suhrakampf, 1972; see also Can long waves of capitalist development be explained?, 'Futures' August 1981; The heyday of capitalism and its aftermath, 'Socialist Register' 1964; 'Long Waves of Capitalist Development' Cambridge, 1980.
12. L.Z. Pasinetti, 'Structural Change and Economic Growth: a theoretical essay on the dynamics of the wealth of nations' Cambridge, 1981.
13. J.A. Schumpeter, 'Business Cycles: A Theoretical, Historical and Statistical Analysis of the Capitalist Process' 2 vols, NY, McGraw-Hill, 1939, p.666.
14. J. Kuczynski, 'Special Analysis and Cluster Analysis Mathematical Methods' paper prepared for 7th Intl Congress on Economics, Edinburgh, 1978, p.81.

15. I. Wallerstein, 'The Capitalist World Economy' Cambridge and New York, Cambridge University Press, 1979.
16. A.G. Frank, 'Low Profit Invention and High Profit Innovation in Technological Change' Research Memorandum No. 8218, University of Amsterdam, 1983.
17. S. Gomulka, 'Inventive Activity, Diffusion and the Stages of Economic Growth' Skrifter fra Aarhus Univeerstets Okonomiske Institut no.24, 1971.
18. Op.cit., pp.140, 141.
19. W.W. Rostow, 'The Stages of Economic Growth' London, Cambridge University Press, 1960.
20. S. Davies, 'The Diffusion of Process Innovations' Cambridge, Cambridge University Press, 1979, p.32.
21. Ibid., p.34.
22. Ibid., chapter 4.
23. J.S. Metcalfe, Impulse and diffusion in the study of technical change, 'Futures' October 1981; 'On the diffusion of innovation and the evolution of technology' paper prepared for the TCC Conference, London, January 1982.
24. G. Dosi, Technological paradigms and technological trajectories - a suggested interpreetation of the determinants and directions of technical change, 'Research Policy' vol.II, no.3, 1982, pp.147-164.
25. This is probably one of the most significant factors in the slow and uneven inter-county diffusion of steel technology in the 1960s and 1970s. See in particular Gold et al.'s analysis of the diffusion of steel technology in the US: B. Gold, W.S. Pierce and G. Rosegger, Diffusion of major technological innovations in US and steel manufacturing, 'Journal of Industrial Economics, vol.18, 1970, pp.218-41; and Diffusion of major technological innovations, in B. Gold (ed.), 'Technological Change, Economics, Management and Environment' Oxford, Pergamon, 1975.
'The slow diffusion of the open hearth (even) in the face of sharply growing output was clearly attributable to the industry's heavy financial and technical commitments only a few years earlier to the Bessemer process.' (Gold, op.cit. p.139)
In a similar vein, L. Nabseth and G. Ray ('The Diffusion of New Industrial Processes, Cambridge, Cambridge University Press, 1974) noted that (p.303):
'It could be argued that even the most farsighted management could not have foreseen, in the early 1960s when many oxygen steel decisions were being made, the drastic changes and improvements in continuous casting that were to occur in the late 1960s.' 26. G. Ray, Innovation in the long cycle, 'Lloyds Bank Review' no. 135,January 1980, pp.14-28.
27. Ibid., p.20.
28. L. Soete, Technical change, catching up and the productivity slowdown, in O. Granstrand and J. Sigurdson (eds), 'Technological and Industrial Policy in China and Europe' proceedings of the first joint TIPCE conference, 1981, RPI, University of Lund, pp.96-115.
29. Following the probit model discussed above, we assume for the sake of the present argument that only these countries have achieved the necessary critical income level to enable them to assimilate and adopt foreign technology, and use effectively the international technology diffusion growth potential.
30. L. Soete and G. Dosi, 'Technology and Employment Opportunities in the Electrical and Electronics Industries' 1983.
31. M. Teubal, The R and D Performance Through Time of Young High-Technology Firms, 'Research Policy' XI, 1983.
32. J. Surrey and W. Walker, 'The European Power Plant Industry: Structural Responses to International Market Pressures' Sussex European Papers no.12, Sussex European Research Centre, University of Sussex, 1981.

13 The role of small firms in the emergence of new technologies

Roy Rothwell

It is clear from recent policy statements on technological and economic change that, generally speaking, governments in the advanced market economies increasingly have laid greater emphasis on measures to support small and medium sized manufacturing firms (Rothwell and Zegveld, 1981). (In Europe, for the purposes of government policy, this generally means firms with employment of between 1 and 499.) This is based on the belief that small and medium sized firms (SMFs) are a potent vehicle for the creation of new jobs, for regional economic regeneration and for enhancing national rates of technological innovation.

The debate concerning firm size and innovation is of long standing, some commentators arguing that large size and monopoly power are prerequisites for economic progress via technological change, others that because of behavioural and organizational factors small firms are better adapted to the creation of major innovations. Some of the advantages and disadvantages variously ascribed to large and small firms in innovation are listed in Table 1, which suggests apriori that comparative advantage in innovation is unequivocally associated neither with large nor small scale.

Data from the UK on some 2,300 important innovations introduced by - though not necessarily developed by - British companies during the period 1945 to 1980 have thrown some light on this issue (Townsend et al., 1981). These data (on some 35 sectors of industry) showed that, at an aggregate level, SMFs' share of innovations in the UK has, during the period covered, consistently averaged about 20% of the total (Table 2). At the same time, the share enjoyed by firms in the largest size category (greater than 10,000 employees) increased progressively from 36% during 1945 to 59% during 1975-80. Moreover, the data also showed that the larger firms increasingly have innovated via smaller units and that independent firms increasingly were displaced by subsidiaries of larger firms as the major source of innovations (Table 3).

Simply counting innovations, of course, tells us nothing about the relative innovative efficiency of small and large firms (nor about the degree of 'radicalness' of the different innovations) measured as

Table 1. Advantages and disadvantages of small and large firms in innovation (statements in parentheses represent areas of potential disadvantage).

	Small Firms
Marketing	Ability to react quickly to keep abreast of fast-changing market requirements. (Market start-up abroad can be prohibitively costly.)
Management	Lack of bureaucracy. Dynamic, entrepreneurial managers react quickly to take advantage of new opportunities and are willing to accept risk.
Internal Communication	Efficient and informal internal communication networks. Affords a fast response to internal problem solving; provides ability to reorganize rapidly to adapt to change in external environment.
Qualified Technical Manpower	(Often lack suitably qualified technical specialists. Often unable to support a formal R&D effort on an appreciable scale.)
External Communication	(Often lack the time or resources to identify and use important external sources of scientific and technological expertise.)
Finance	(Can experience great difficulty in attracting capital, especially risk capital. Innovation can represent disproportionately large financial risk. Inability to spread risk over a portfolio of projects.)
Economies of scale and the systems approach	(In some areas scale economies form substantial entry barrier to small firms. Inability to offer integrated product lines or systems.)
Growth	(Can experience difficulty in acquiring external capital necessary for rapid growth. Entrepreneurial managers sometimes unable to cope with increasingly complex organizations.)
Patents	(Can experience problems in coping with the patent system. Cannot afford time or costs involved in patent litigation.)
Government Regulations	(Often cannot cope with complex regulations. Unit costs of compliance for small firms often high.)

Source: Rothwell and Zegveld (1982).

Large Firms	
Comprehensive distribution and servicing facilities. High degree of market power with existing products.	Marketing
Professional managers able to control complex organizations and establish corporate strategies. (Can suffer an excess of bureaucracy. Often controlled by accountants who can be risk-averse. Managers can become mere "administrators" who lack dynamism with respect to new long-term opportunities.)	Management
(Internal communications often cumbersome; this can lead to slow reaction to external threats and opportunities.)	Internal Communication
Ability to attract highly skilled technical specialists. Can support the establishment of a large R&D laboratory.	Qualified Technical Manpower
Able to "plug-in" to external sources of scientific and technological expertise. Can afford library and information services. Can subcontract R&D to specialist centres of expertise. Can buy crucial technical information and technology.	External Communication
Ability to borrow on capital market. Ability to spread risk over a portfolio of projects. Better able to fund diversification into new technologies and new markets.	Finance
Ability to gain scale economies in R&D, production and marketing. Ability to offer a range of complementary products. Ability to bid for large turnkey projects.	Economies of scale and the systems approach
Ability to finance expansion of production base. Ability to fund growth via diversification and acquisition.	Growth
Ability to employ patent specialists. Can afford to litigate to defend patents against infringement.	Patents
Ability to fund legal services to cope with complex regulatory requirements. Can spread regulatory costs. Able to fund R&D necessary for compliance.	Government Regulations

innovations per unit of employment or innovations per unit of output. Nor does it inform us of relative R&D efficiency, i.e. innovations per unit of R&D expenditure. In relation to the former measure, Wyatt (1982) has suggested on the basis of the SPRU innovation data that in general the innovative efficiency of the very largest firms consistently has been higher than that of their smaller counterparts during the whole of the 1945-80 period. Regarding R & D efficiency, Wyatt (1982) found that:

> small firms' (employment between 100 and 500 employees) share of innovations is considerably greater than their share of R&D expenditure for all thirty-five industry sectors. This is sometimes interpreted as small firms' greater efficiency in R & D activities. Another explanation is that there is a lower degree of functional specialisation in small firms, so that a higher proportion of innovative activities occurs outside of what is defined as R & D activities.

At a more disaggregated level, relative innovative efficiency varied between sectors and in some sectors (e.g. mining machinery, textile machinery, electronic capital goods and scientific instruments) small firms enjoyed innovative efficiencies greater than unity. In terms of share in total sectoral innovations, in some sectors (e.g. pharmaceuticals) SMFs played a very small or zero role; in other sectors (e.g. scientific instruments) SMFs played a consistently significant role. Not surprisingly, where R&D requirements are very large and capital costs very high, high entry costs prohibit the participation of small firms; where technical, capital and marketing start-up costs are relatively low, entry by small firms is entirely possible.

An interesting case is electronic computers. Between 1945 and 1969, innovative activity (and output) was dominated by large firms producing predominantly mainframe computers. This involved high capital costs, a large R&D effort and the establishment of comprehensive production and servicing facilities. During the period 1970-1980, however, SMFs have emerged as a significant force and accounted for 40% of all important innovations introduced in the UK. This reflects the introduction of high density integrated circuits and the microprocessor, which made possible the entry of new small firms producing mini- and microcomputers. These are skill-intensive, require considerably less capital investment than previous models and have opened up a large variety of new market niches suitable for exploitation by technical entrepreneurs. Thus, while one type of technological change, i.e. that requiring high development costs and large investment for commercial realization, can pose a barrier to entry by small firms, other types of technological change can provide them with many new opportunities.

Table 2. Percentage of innovations in each firm size category for each five-year period.

No. of employees	1945-49 %	1950-54 %	1955-59 %	1960-64 %	1965-69 %	1970-74 %	1975-80 %	Total %
1-199	16.0	12.0	11.0	11.0	13.0	15.0	17.0	14.0
200-499	9.0	6.0	6.0	6.0	7.0	9.0	7.0	7.0
500-999	3.0	2.0	7.0	5.0	5.0	4.0	3.0	4.0
1000-9999	36.0	36.0	25.0	27.0	23.0	17.0	14.0	23.0
10,000 and over	36.0	44.0	50.0	51.0	52.0	55.1	59.0	52.0
Total	100.0	100.0	100.0	100.0	100.0	100.0	100.0	100.0
No. of innovations	94	191	274	405	467	401	461	2293

Source: S. Wyatt (1982), "The Role of Small Firms in Innovative Activity: Some New Evidence", SPRU, Sussex.

Table 3. Percentage of innovations in each unit size category for each five-year period.

No. of employees	1945-49 %	1950-54 %	1955-59 %	1960-64 %	1965-69 %	1970-74 %	1975-80 %	Total %
1-199	20.0	23.0	17.0	14.0	21.0	27.0	32.0	23.0
200-499	11.0	18.0	14.0	14.0	16.0	16.0	14.0	15.0
500-999	9.0	6.0	13.5	12.0	11.0	13.0	16.0	12.0
1000-9999	48.0	41.0	38.0	43.0	40.0	34.0	30.0	38.0
10,000 and over	13.0	12.0	18.0	17.0	12.0	10.0	8.0	13.0
Total	100.0	100.0	100.0	100.0	100.0	100.0	100.0	100.0
No. of innovations	94	191	274	405	467	401	461	2293

Source: S. Wyatt (1982), "The Role of Small Firms in Innovative Activity: Some New Evidence", SPRU, Sussex.

Small firms in the emergence of new technologies

Two important points emerge from the above discussion. The first is that the debate concerning firm size and innovation should proceed on a sector-by-sector basis. The second is that a dynamic approach clearly is necessary: the relative contribution of firms of different sizes to innovation in a particular industry might depend on the age of that industry; the type of innovation typically produced by large and small sized firms at different stages in the industry cycle might also vary, i.e. product or process innovations.

It was, of course, Schumpeter (1939) who made the major initial contribution to the discussion of industrial dynamics. He emphasized that while entrepreneurs play a significant role in the establishment of new industrial branches, during the latter phases of development, innovation requires large firms because of the high costs involved and considerable market power if innovation is to be worthwhile.

According to Freeman et al. (1982), there are essentially two Schumpeterian models of innovation: 'entrepreneurial' innovation and 'managed' innovation. In the first case, new basic technologies emerge which are coupled in an unspecified way to new scientific developments and which are largely exogenous to existing companies and market structures. Risk-taking entrepreneurs grasp the techno-economic opportunities thus offered and, via radical innovation, foster the growth of new industries and the emergence of major new product groups. It is during this phase of the industry cycle that dynamic new, small but fast growing firms play the key role as innovators.

As technology and markets mature and average firm size increases, inventive activity becomes progressively internalized in the form of large in-house research and development laboratories (which remain coupled to external sources of science and technology). As Freeman et al. put it:

> In Schumpeter (model) II therefore there is a strong positive feedback loop from successful innovation to increased R & D activities setting up a 'virtuous' self-reinforcing circle leading to renewed impulses to increased market concentration. Schumpeter now sees inventive activites as increasingly under the control of large firms and reinforcing their competitive position. The 'coupling' between science, technology, innovative investment and the market, once loose and subject to long delays, is now much more intimate and continuous.

Finally, as the industry and its technology mature, the possibilities for major product innovations diminish. At the same time, market requirements become increasingly well specified and competing products are little differentiated technically. As a result, price becomes a more significant factor in competition, and development efforts become more and more directed towards process efficiency improvement (cost reduction). If this pattern of evolution is valid, then while the initial small entrepreneurial firms are concerned primarily with new product innovation and major product improvement, the subsequent large established firms become increasingly involved in process innovation and minor product improvements. A final point is that the debate so far has focussed largely on the issue of small firms or large firms and generally has failed to recognise that the two will often be related. In other words, we should be concerned with the 'dynamic complementarities' that can exist between small and large firms during the industry cycle. Below we shall attempt to illustrate this process and, in particular, point to the important role new-technology-based firms (NTBFs) can play during the early phases of industrial evolution.

The evolution of the US semiconductor industry

An approximate example of Schumpeterian industrial evolution, and one which illustrates the importance of large firm/small firm complementarities, can be found in the evolution of the US semiconductor industry. The following description is taken largely from Rothwell and Zegveld (1982).

The beginning of the semiconductor industry can be traced to the invention of the transistor effect at Bell Telephone Laboratories in 1947 by Bardeen and Brattain. Although their findings paved the way for the invention of the bipolar junction transistor, the real breakthrough came in 1952 when Shockley, the research team leader, described a field effect transistor with a central electrode consisting of a reverse-biased junction. Shockley subsequently left Bell Labs and several years later he established his own company in his native Palo Alto, backed by finance from the Clevite Corporation. Shockley attracted a number of leading physicists and engineers into his company but, in 1957, eight of his brightest people left to form their own company. This marked the beginning of the rapid growth of NTBFs in the Palo Alto area, which subsequently gave it the name of Silicon Valley. While a number of other centres of semiconductor production were emerging concurrently, notably at Dallas, Texas (Texas Instruments) and Phoenix, Arizona (Motorola), nevertheless Silicon Valley has been exceptional in world terms in the amount of semiconductor production and technological innovation that has occurred in such a concentrated area.

The eight ex-Shockley workers obtained backing from the Fairchild Camera Corporation, which had actively been seeking diversification, and in September 1957 Fairchild Semiconductor was founded in Mountain View, California. In 1959, the Fairchild Camera Corporation exercised an option to buy a majority interest in Fairchild Semiconductor. The latter grew rapidly from sales of $0.5 million in 1960 to $27 million in 1967 to $520 million in 1978. During the next few

Small firms in the emergence of new technologies

years there was considerable spin-off from Fairchild Semiconductor of both people and technology, and many companies were formed by people formerly with, or associated with, Fairchild. This process has been described by Mason (1979):

> The first spin-off was in 1959, when Baldwin, not from the original Shockley team, left Fairchild to form Rheem Semiconductor, collecting on the way people from Hughes Aircraft. In 1961, four of the originals left to form Amelco and one of these, Hoeni, left in 1964 to form Union Carbide Electronics, moving on in 1967 to form Intersil. Of ... interest ... was another event in 1961, when Signetics was formed. This was formed by four people who were a significant part of the Fairchild Semiconductor team... They managed to get venture capital backing from the Dow-Corning group for this move.

At the same time that new-technology-based small firms were being spawned in Silicon Valley, Bell Labs (a subsidiary of AT&T) continued with its vigorous inventive and innovative activity, although all AT&T's output (via Western Electric) was produced for its own use in order to avoid anti-trust litigation. Bell Labs, along with other major companies, have between them accounted for a high percentage of all major innovations in semiconductor technology, which is illustrated in Table 4 for the two decades up to 1971. Interestingly, since 1976, major Japanese companies have made an increasing contribution to technological advance in semiconductors: Sharp's automatic bonding on 'exotic' substrates in 1977; Mitsubishi's vertical injection logic and V-MOS in 1978; Fujitsu's 64K chip in 1978 (Dosi, 1981).

Despite the initial dominance of large companies in basic invention in the semiconductor field, NTB small firms played a key role in commercial exploitation, especially during the earlier stages in the US semiconductor industry's development. What in fact occurred during the evolution of the US semiconductor industry was a classical example of the 'dynamic complementarities' that can exist between large and small firms. Existing large firms provided much of the basic technology, venture capital and technically skilled personnel which were essential to NTBF start-ups; the NTBFs provided the risk-taking entrepreneurial drive and rapid market exploitation. It was a synergistic relationship.

From the late 1960s onwards, the output of the US semiconductor industry began increasingly to be concentrated in the top ten or so companies. Production economies of scale grew in importance (and plant size increased), as did production learning, and firms began actively to seek rapid movement down the production learning curve. The importance of price in competition increased as the unit cost of semiconductor component production decreased. According to

Table 4. Major product and process innovations in the US semiconductor industry

Innovation	Principal company responsible	Date
Point contact transistor	Bell Telephone Laboratories	1951
Grown junction transis.	Bell Telephone Laboratories	1951
Alloy junction transis.	General Electric Co., RCA Corp.	1952
Surface barrier transis.	Philco Corp.	1954
Silicon junction trans.	Texas Instruments Inc.	1954
Diffused transistor	Bell Telephone Laboratories Texas Instruments Inc.	1956
Silicon controlled rectifier	General Electric Co.	1956
Tunnel diode	Sony (Japan)	1957
Planar transistor	Fairchild Camera & Instrument Corp.	1960
Epitaxial transistor	Bell Telephone Laboratories	1960
Integrated circuit	Texas Instruments Inc. Fairchild Camera & Instrument Corp.	1961
MOS transistor	Fairchild Camera & Instrument Corp.	1962
DTL integrated circuit	Signetics Corp.	1962
ECL integrated circuit	Pacific (TRW)	1962
Gunn diode	Intl Business Machines Corp.	1963
Beam lead	Bell Telephone Laboratories	1964
TTL integrated circuit	Pacific (TRW)	1964
Light-emitting diode	Texas Instruments Inc.	1964
MOSFET (MOS field effect)	Bell Telephone Laboratories Philips (Holland)	1968
Collector diffusion isolation	Bell Telephone Laboratories	1969
Schottky TTL	Texas Instruments Inc.	1969
Charge-coupled device	Bell Telephone Labs, Fairchild Camera	1969
Complementary MOS	RCA Corp.	1969
Silicon-on-sapphire	RCA Corp.	1970
Ion implementation	Bell Telephone Laboratories	1971

Source: D.W. Webbink, "The Semiconductor Industry: Structure, Conduct and Performance", unpublished staff report to the US Federal Trade Commission, January 1977.

Sciberras (1977), the prime motive for rapid cost reductions was to deter new entrants by creating significant scale barriers to entry in addition to technological entry barriers. This might at least partially explain why semiconductor technology was exploited in Europe mainly by large existing electronics companies: Europe entered the race at a late date, by which time existing scale and technological barriers largely precluded entry by new small firms. (Also, military procurement had played a key role in the high rate of NTBF formation in the US.)

Earlier we implied the possibility of a move from a focus largely on productive innovation to one largely on process innovation, as an industry and its technology mature. Figure 1 plots the cumulative number of patents issued in the US in the areas of 'semiconductor internal structure technology' and 'semiconductor preparation technology' between 1963 and 1974. It indicates that the balance of inventive activity has indeed shifted from product (internal structure technology) to process (semiconductor preparation technology), which might be taken as support for the validity of our argument.

Thus, in the development of the US semiconductor industry, we see an example of Schumpeterian industrial evolution from the 'entrepreneurial' model of the newly emerging industry, to the 'managed' model of the mature international oligopoly of today. Nowadays, the main opportunities for new entrants appear to be not in semiconductor production itself, but rather in the application of semiconductor devices to the production of new products, notably in the general area of information technology, currently mooted as the new industry of the next decade. Indeed, as Table 5 indicates, in the US it is mainly young firms in the information technology field which are currently enjoying the fastest rates of growth in sales, with accompanying high employment growth rates. This represents no mean feat during a period of world recession. At the same time, existing large companies such as IBM and Exxon are moving into the information technology field, the latter through the acquisition of innovative companies in this area.

The evolution of the computer- aided design industry

A second example of industrial evolution that indicates an important role for NTBFs can be found in the case of the computer aided design (CAD) industry. The data below are drawn from the work of Kaplinsky (1981, 1982), who has identified four main phases in the development of CAD: pre-1969 (industry origins); 1969-1974 (dynamic new firms); 1974-1980 (the trend to concentration); post-1980 (maturity).

During the first phase, development was concentrated in established large companies in the defence, aerospace and aeronautical industries in collaboration with mainframe computer manufacturers. In the late

Figure 1

Number of patents issued in the USA
in Semiconductor Structure Technology
and Semiconductor Preparation Technology

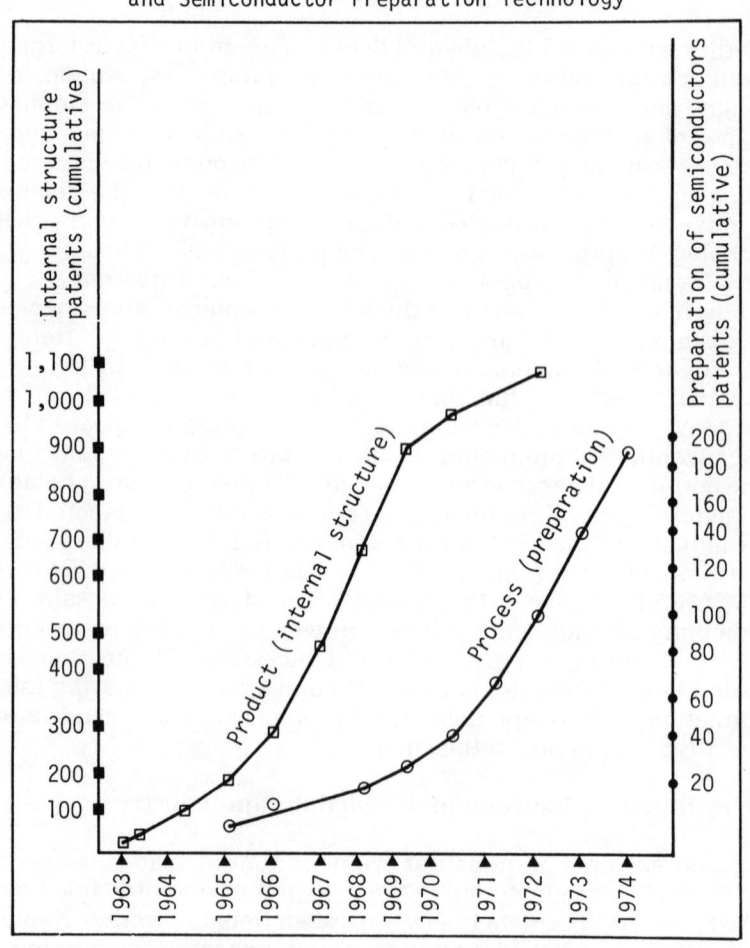

Source: Rothwell and Zegveld, 1982.

Table 5. US firms with annual average growth rates (sales 1976-80) of more than 40%.

Firm	Sector	1976-80 av.annual growth			1980		
		Sales %	Profits %	Empl. %	Sales (million$)	R&D/ sales %	Profits/ sales ratio(%)
Tandem Computers	IP*:computers	247.4	284.5	n.a.	109	8.1	10.1
Cray Research	IP :computers	197.0	126.8	71.8	61	14.5	18.0
Apple Computer**	IP :computers	144.7	n.a.	11.2	117	6.2	10.3
Floating Point Systems	IP :computers	120.5	67.0	n.a.	42	10.9	9.5
Intermedics	Electronics	111.5	113.4	53.3	105	4.6	10.5
Triad Systems	IP:peripherals serv.	89.5	110.3	n.a.	57	6.6	8.8
Prime Computers	IP: computers	88.1	132.0	73.8	268	7.6	11.6
Rolm	Telecommunications	79.0	101.4	69.7	201	6.7	8.5
Lamson & Sessions	Misc.manufacturing	65.3	n.a.	26.8	235	0.6	-1.3
Auto-Trol Technology	IP:peripherals serv.	64.1	96.9	n.a.	51	12.1	7.8
Data Terminal Systems	IP:peripherals serv.	64.0	n.a.	49.5	118	5.1	-2.5
Computervision	IP:peripherals serv.	59.0	116.3	44.7	224	9.9	10.3
Paradyne	Electronics	57.9	132.8	51.8	76	8.4	10.5
Siltec	Electronics	56.5	64.0	n.a.	57	3.6	5.3
Advanced Micro Devices	Semiconductors	56.2	91.0	48.6	226	12.5	10.2
Savin	IP:office equipment	53.7	90.2	33.5	357	2.3	7.8
American Mgmt Syst.	IP:peripherals serv.	53.3	57.2	n.a.	59	7.1	3.4
CPT	IP:office equipment	49.4	55.5	37.9	59	3.4	10.2
Wang Laboratories	IP:office equipment	48.5	72.1	n.a.	543	6.7	9.6
Storage Technology	IP:peripherals serv.	48.2	58.0	n.a.	604	6.5	7.5
Comshare	IP:peripherals serv.	47.6	51.4	38.9	78	6.3	5.1
Datapoint	IP:peripherals serv.	47.3	61.8	36.6	50	8.7	10.3
Verbatim	IP:peripherals serv.	47.3	28.5	45.6	319	5.8	2.0
US Surgical Instru.	Instruments	46.9	52.8	34.1	86	3.5	9.3
Sega Enterprises	Misc.manufacturing	46.8	55.5	9.3	140	1.2	8.6
Parker Pen	Misc.manufacturing	45.2	34.8	11.2	664	0.4	6.0
TIE Communications	Electronics	44.9	94.3	n.a.	60	2.5	5.0
Kratos	Instruments	44.0	44.5	n.a.	56	7.5	5.4
Intel	Semiconductors	43.9	43.4	27.5	855	11.3	11.3
Datacard	IP:peripherals serv.	42.6	55.5	n.a.	66	2.6	10.6
Gerber Scientific	Instruments	41.6	83.9	32.1	74	7.1	8.1
Data General	IP:computers	41.0	33.0	31.0	654	10.0	8.4
Computer Consoles	IP:peripherals serv.	40.7	55.8	16.1	44	10.5	11.4
Analogic	Electronics	40.3	67.7	n.a.	67	8.9	9.0
Miller (Herman)	Misc.manufacturing	40.3	40.0	35.1	230	2.5	5.2
Pengo Industries	Machinery	40.1	22.8	n.a.	78	1.9	3.8

*IP = Information processing
**For Apple Computer Inc. which only became publicly held in 1980, the growth in sales relates only to the period 1979-80.

Source: Freeman, Clark and Soete (1982). Calculated from "Business Week", 6 June 1981.

1960s, General Motors entered the field with the development of its 'Design Augmented by Computers' programme.

> In summary, therefore, during this early period there was hardly any 'market' for CAD, with most developments occurring to assist own-use by large, technically advanced engineering corporations in the US and (to a lesser extent) in the UK. (Kaplinsky, 1981)

The second phase was characterized by the emergence of new small spin-off firms in the US (from both CAD producers and electronics companies), which played the primary role in the rapid diffusion of CAD devices into the electronics industry. Several of these firms grew extremely rapidly to become, along with IBM, today's market leaders. In Europe, by contrast, the major existing electronics firms developed CAD equipment for their own use.

> In summary, therefore, this second period of industry development saw the emergence of new, independent firms and the rapid diffusion of the technology out of the defence, aerospace and automobile sectors to the electronics sectors. (Kaplinsky, 1981)

The third phase saw the rapid diffusion in use of CAD across manufacturing, a process in which the 'newcomers' played a key role. During this period of extremely rapid market growth, the industry become increasingly concentrated, 93% of the US market share in 1980 being held by eight companies (notably Computervision with 33.2% of the total). At the same time, patterns of ownership began to change and there was a series of takeovers by major corporations of several of the fast-growing newcomers.

> To summarize, therefore, this third phase of industry development was associated with the growing size of CAD firms, the growing organic trend towards concentration within the sector, and a tendency for formerly independent CAD firms to be swallowed by existing trans-national corporations. (Kaplinsky, 1981)

At the beginning of the current phase of development, the market was dominated by turnkey suppliers supplying either mainframe systems (user entry costs of about $500,000) or minicomputer systems (user entry costs of about $200,000). From 1980 onwards, as the user base was broadened, a market niche has emerged for dedicated systems. These are based not on a comprehensive and flexible package of software applications, but on limited software packages for specific applications. A number of microcomputer-based companies, founded by spin-offs from existing CAD suppliers, have begun to emerge, offering systems for as little as $30,000 each.

> To summarize, therefore, this most recent stage of industry development has seen two divergent trends - a continued tendency to concentration and an opposing tendency for the entry of new small firms selling limited capability dedicated systems. (Kaplinsky, 1981)

In the UK, it has been estimated that US firms held a 62% share of all CAD systems installed up to mid-1981 (Arnold, 1982). Of the remaining 38% share of installations, 17% were held by subsidiaries of large electronic companies established in the late 1960s, 12% by spin-off companies, 5% by a public body (essentially a software house) and 4% by other companies.

Discussion

From the above brief descriptions of the evolution of two high-technology industries we can draw out a number of significant factors:

* While established large US corporations played an important early role in invention both in semiconductors and CAD technology, NTBFs played a key role in innovation and diffusion. In both instances, the early inventive activity was geared towards 'own use'
* In the case of semiconductors, much of the dynamic growth and market diffusion came about as a result of the formation and rapid expansion of NTBFs.
* In the CAD industry, NTBFs played the key role in the rapid diffusion of CAD systems to electronics and other areas.
* In both cases, the technological entrepreneurs often came from established corporations, bringing a great deal of technological and market know-how with them.
* In both cases, established corporations and venture capital institutions played an important part in funding the start-up and growth of NTBFs.
* In both cases the industries became highly concentrated and subject to external takeover.
* In both cases, as the industries matured, scale economies became increasingly important in the mainstream activities and strong oligopolies were formed, leaving on specialist market niches for new and small suppliers.

It is clear from our description of the evolution of the semiconductor and CAD industries that it is indeed necessary to consider the interactions between small and large firms if we are fully to understand the evolutionary dynamic. In both instances, existing large corporations played the major initial role in invention, producing new devices largely for in-house use only. The major role in the initial market diffusion of these new devices, however, was played by new, small but fast-growing companies founded by technological entrepreneurs. Moreover, the technical know-how, the venture capital and the entrepreneurs themselves very often derived from the established corporations, as well as, in the case of the latter two, from major companies operating in other areas. Thus we see - at least in the US - a system of dynamic complementarity between the large and the small: both had their unique contribution to make; both were necessary to the initiation of the new techno/market combinations and to their rapid commercial exploitation.

What our discussions do suggest is that established technology-based large corporations can be extremely effective in creating new technological possibilities; they are highly inventive. While they are adept at utilizing the results of their inventiveness in-house (new technology for existing applications), they are less well adapted to the rapid exploitation of their inventions in new markets (new technology for new applications). It appears that new firms, initially, are better adapted to exploit new techno/market combinations; they are unconstrained by existing techno/market regimes within which established corporations might be rather strongly bound. Referring back to Table 1, it appears that during the early phases in the evolution of a new industry, the behavioural advantages of small scale are crucial; as the industry evolves, technological possibilities become better defined and market needs become increasingly well specified, and the advantages of large scale begin to dominate. Comparative advantage shifts to the larger firms, and the industry develops towards a mature oligopoly - a situation characteristic of the semiconductor and CAD industries today.

The question now, of course, is what about today: can we detect indications of technological entrepreneurship in areas of great growth potential for the coming decades? We have already dealt with CAD and the role of NTBFs in its rapid growth phase, but it is worth adding that markets of up to $12 billion for CAD systems have been predicted for 1990 (Kaplinsky, 1982). In the general area of information processing, as Table 5 indicated, out of 36 companies which enjoyed an annual growth rate in sales in the US between 1976 and 1980 of greater than 40%, 23 (63%) were in the information processing area. Between them, these relatively young firms had combined sales in 1980 of approximately $4,000 million, despite a period of deepening world recession.

Turning finally to biotechnology, the economic potential of this infant industry is immense, and many new firms have been established within this area - most notably in the USA - during recent years. It is interesting that, when it became generally realized that biotechnology is still in its research-intensive phase (the commercialization of biotechnological products on a large scale being a thing of the late 1980s onwards), private venture capitalists began to have second thoughts concerning their investment in the newcomers. Increasingly, established large corporations stepped in to fill the venture capital gap. In fact, we can today see synergistic interactions occurring between a number of large firm/small firm combinations. Dow Chemicals and Monsanto, for example, while pursuing their own research and development programmes, are also investing in smaller companies; Biotechnology General, an entrepreneurial newcomer in the USA, is negotiating for venture capital with three large firms to help finance the development of three new agricultural products; Bio Isolates of Swansea is now setting up a joint venture with Dunlop; Grand Metropolitan has invested more than 4 million in Biogen, a new Swiss-based biotechnology company; and so on. Whether or not any of the new small biotechnology firms become the giants of the future is, today, less important than their role in stimulating widespread interest and greatly increased investment in biotechnological research and development.

We can thus see that NTBFs, by themselves, are no panacea for rapid growth and economic recovery. The vigorous innovatory efforts of many NTBFs in parallel with, and in many cases coupled to, the efforts of established major corporations, however, will, together, greatly increase the possibility of the world economy innovating itself into the recovery phase of the fifth Kondratiev. Both are desirable; both are essential.

Finally, we turn briefly to Japan, where a handful of dominant, major corporations have demonstrated great corporate flexibility and technological innovativeness during the past 30 or so years. Even here, however, small firms have played a key role by providing the major corporations with an effective source of low-cost labour, the ability to specialize in the capital-intensive final phases of production, great employment flexibility and financing flexibility (Twaalfhoven and Hattori, 1982). Indeed, in Japan in 1978, small businesses (employment below 300) accounted for 53% of the value of all manufacturing shipments. More recently, MITI has shown an interest in stimulating the creation of independent new enterprises and in 1982 announced plans to increase the availability of venture capital. While this move owes much to growing concern over employment loss in dependent subcontractors, it also owes something to a growing desire in Japan to increase the national capacity for creativity and inventiveness. In other words, the potential of independent small enterprises for technological entrepreneurship is now gaining increased recognition in Japan.

References

E. Arnold (1982) 'Competition and Policy in a Knowledge-Intensive Industry: CAD Equipment Supply in the UK' SPRU, Sussex, July.
G. Dosi (1981) Institutions and Markets in High Technology Industries: An Assessment of Government Intervention in European Microelectronics' in C.F. Carter (ed), 'Industrial Policies and Innovation' Heinemann, London.
C. Freeman, J. Clark and L. Soete (1982) 'Unemployment and Technical Innovation' Frances Pinter, London.
R. Kaplinsky (1981) Firm Size and Technical Change in a Dynamic Context, 'Journal of Industrial Economics' Institute of Development Studies, Sussex.
R. Kaplinsky (1982) 'The Impact of Technical Change on the International Division of Labour: The Illustrative Case of CAD' Frances Pinter, London.
D. Mason (1979) 'Factors Affecting the Successful Development and Marketing of Innovative Semiconductor Devices' PhD thesis, PCL, London and SPRU, Sussex.
R. Rothwell (1983) Firm Size and Innovation: A Case of Dynamic Complementarity, 'Journal of General Management' Spring.
R. Rothwell and W. Zegveld (1981) 'Industrial Innovation and Public Policy' Frances Pinter, London.
R. Rothwell and W. Zegveld (1982) 'Innovation and the Small and Medium Sized Firm' Frances Pinter, London.
J.A. Schumpeter (1939) 'Business Cycles' McGraw-Hill, New York and London.
E. Sciberras (1977) 'Multinational Electronic Companies and National Economic Policies' JAI Press, Greenwich, Conn.
J. Townsend, F. Henwood, G. Thomas, K. Pavitt and S. Wyatt (1981) 'Innovations in Britain Since 1945' SPRU Occasional Paper Series No.16, Sussex.
F. Twaalfhoven and T. Hatteri (1982) 'The Supporting Role of the Small Japanese Enterprise' Indivers Research, Netherlands, October.
S. Wyatt (1982) 'The Role of Small Firms in Innovative Activity: Some New Evidence' SPRU, Sussex, May.